Artificial Light

M. Luckiesh

I0484218

DEDICATED

TO THOSE WHO HAVE ENCOURAGED ORGANIZED SCIENTIFIC RESEARCH FOR THE ADVANCEMENT OF CIVILIZATION

PREFACE

In the following pages I have endeavored to discuss artificial light for the general reader, in a manner as devoid as possible of intricate details. The early chapters deal particularly with primitive artificial light and their contents are generally historical. The science of light-production may be considered to have been born in the latter part of the eighteenth century and beginning with that period a few chapters treat of the development of artificial light up to the present time. Until the middle of the nineteenth century mere light was available, but as the century progressed, the light-sources through the application of science became more powerful and efficient. Gradually mere light grew to more light and in the dawn of the twentieth century adequate light became available. In a single century, after the development of artificial light began in earnest, the efficiency of light-production increased fifty-fold and the cost diminished correspondingly. The next group of chapters deals with various economic influences of artificial light and with some of the byways in which artificial light is serving mankind. On passing through the spectacular aspects of lighting we finally emerge into the esthetics of light and lighting.

The aim has been to show that artificial light has become intricately interwoven with human activities and that it has been a powerful influence upon the progress of civilization. The subject is too extensive to be treated in detail in a single volume, but an effort has been made to present a discussion fairly complete in scope. It is hoped that the reader will gain a greater appreciation of artificial light as an economic factor, as an artistic medium, and as a mighty influence upon the safety, efficiency, health, happiness, and general progress of mankind.

M. LUCKIESH.

CONTENTS

CHAPTER

I LIGHT AND PROGRESS

II THE ART OF MAKING FIRE

III PRIMITIVE LIGHT-SOURCES

IV THE CEREMONIAL USE OF LIGHT

V OIL-LAMPS OF THE NINETEENTH CENTURY

VI EARLY GAS-LIGHTING

VII THE SCIENCE OF LIGHT-PRODUCTION

VIII MODERN GAS-LIGHTING

IX THE ELECTRIC ARCS

X THE ELECTRIC INCANDESCENT FILAMENT LAMPS

XI THE LIGHT OF THE FUTURE

XII LIGHTING THE STREETS

XIII LIGHTHOUSES

XIV ARTIFICIAL LIGHT IN WARFARE

XV SIGNALING

XVI THE COST OF LIGHT

XVII LIGHT AND SAFETY

XVIII THE COST OF LIVING

XIX ARTIFICIAL LIGHT AND CHEMISTRY

XX LIGHT AND HEALTH

XXI MODIFYING ARTIFICIAL LIGHT

XXII SPECTACULAR LIGHTING

XXIII THE EXPRESSIVENESS OF LIGHT

XXIV LIGHTING THE HOME

XXV LIGHTING--A FINE ART?

ARTIFICIAL LIGHT

I

LIGHT AND PROGRESS

The human race was born in slavery, totally subservient to nature. The earliest primitive beings feasted or starved according to nature's bounty and sweltered or shivered according to the weather. When night fell they sought shelter with animal instinct, for not only were activities almost completely curtailed by darkness but beyond its screen lurked many dangers. It is interesting to philosophize upon a distinction between a human being and the animal just below him in the scale, but it may serve the present purpose to distinguish the human being as that animal in whom there is an unquenchable and insatiable desire for independence. The effort to escape from the bondage of nature is not solely a human instinct; animals burrow or build retreats through the instinct of self-preservation. But this instinct in animals is soon satisfied, whereas in human beings it has been leading ever onward toward complete emancipation.

The progress of civilization is a long chain of countless achievements each one of which has increased man's independence. Early man perhaps did not conceive the idea of fire and then set out to produce it. His infant mind did not operate in this manner. But when he accidentally struck a spark, produced fire by friction, or discovered it in some other manner, he saw its possibility. It is thrilling to picture primitive man at his first bonfire, enjoying the warmth, or at least interested in it. But how wonderful it must have become as twilight's curtain was drawn across the heavens! This controllable fire emitted light. It is easy to imagine primitive man pondering over this phenomenon with his sluggish mind. Doubtless he cautiously picked up a flaming stick and timidly explored the crowding darkness. Perhaps he carried it into his cave and behold! night had retreated from his abode! No longer was it necessary for him to retire to his bed of leaves when daylight failed. The fire not only banished the chill of night but was a power over darkness. Viewed from the standpoint of civilization, its discovery was one of the greatest strides along the highway of human progress. The activities of man were no longer bounded by sunrise and sunset. The march of civilization had begun.

In the present age of abundant artificial light, with its manifold light-sources and accessories which have made possible countless applications of light, mankind does not realize the importance of this comfort. Its wonderful convenience and omnipresence have resulted in indifference toward it by mankind in general, notwithstanding the fact that it is essential to man's most important and educative sense. By extinguishing the light and pondering upon his helplessness in the resulting darkness, man may gain an idea of its overwhelming importance. Those unfortunate persons who suffer the terrible calamity of blindness after years of dependence upon sight will testify in heartrending terms to the importance of light. Milton, whose eyesight had failed, laments,

O first created beam and thou great Word "Let there be light," and light was over all, Why am I thus bereaved thy prime decree?

Perhaps only through a similar loss would one fully appreciate the tremendous importance of light to him, but imagination should be capable of convincing him that it is one of the most essential and pleasure-giving phenomena known to mankind.

A retrospective view down the vista of centuries reveals by contrast the complexity with which artificial light is woven into human activities of the present time. Written history fails long before the primitive races are reached, but it is safe to trust the imagination to penetrate the fog of unwritten history and find early man huddled in his cave as daylight wanes. Impelled by the restless spirit of progress, this primitive being grasped the opportunity which fire afforded to extend his activities beyond the boundaries of daylight. The crude art upon the walls of his cave was executed by the flame of a smoking fagot. The fire on the ledge at the entrance to his abode became a symbol of home, as the fire on the hearth has symbolized home and hospitality throughout succeeding ages. The accompanying light and the protection from cold combined to establish the home circle. The ties of mated animals expanded through these influences to the bonds of family. Thus light was woven early into family life and has been throughout the ages a moralizing and civilizing influence. To-day the residence functions as a home mainly under artificial light, for owing to the conditions of living and working, the family group gathers chiefly after daylight has failed.

From the pine knot of primitive man to the wonderfully convenient light-sources of to-day there is a great interval, consisting, as appears retrospectively, of small and simple steps long periods apart. Measured by present standards and achievements, development was slow at first and modern man may be inclined to impatience as he views the history of light and human progress. But the achievements of early centuries, which appear so simple at the present time, were really great accomplishments when considered in the light of the knowledge of those remote periods. Science as it exists to-day is founded upon proved facts. The scientist, equipped with a knowledge of physical and chemical laws, is led by his imagination into the darkness of the unexplored unknown. This knowledge illuminates the pathway so that hypotheses are intelligently formed. These evolve into theories which are gradually altered to fit the accumulating facts, for along the battle area of progress there are innumerable scouting-parties gaining secrets from nature. These are supported by individuals and by groups, who verify, amplify, and organize the facts, and they in turn are followed by inventors who apply them. Liaison is maintained at all points, but the attack varies from time to time. It may be intense at certain places and other sectors may be quiet for a time. There are occasional reverses, but the whole line in general progresses. Each year witnesses the acquirement of new territory. It is seen that through the centuries there is an ever-growing momentum as knowledge, efficiency, and organization increase the strength of this invading army of scientists and inventors.

The burning fagot rescued mankind from the shackles of darkness, and the grease-lamp and tallow-candle have done their part. Progress was slow in those early centuries because the great minds of those ages philosophized without a basis of established facts: scientific progress resulted more from an accumulation of accidental discoveries than by a directed attack of philosophy supported by the facts established by experiment. It was not until comparatively recent times, at most three centuries ago, that the great intellects turned to systematically organized scientific research. Such men as Newton laid the foundation for the tremendous strides of to-day. The store of facts increased and as the attitude changed from philosophizing to investigating, the organized knowledge grew apace. All of this paved the way for the momentous successes of the present time.

The end is not in sight and perhaps never will be. The unexplored region extends to infinity and, judged by the past, the momentum of discovery will continue to increase for ages to come, unless the human race decays through the comfort and ease gained from utilizing the magic secrets which are constantly being wrested from nature. Among the achievements of science and invention, the production and application of artificial light ranks high. As an influence upon civilization, no single achievement surpasses it.

Without artificial light, mankind would be comparatively inactive about one half its lifetime. To-day it has been fairly well established that the human organism can flourish on eight hours' sleep in a period of twenty-four hours. Another eight hours spent in work should settle man's obligation to the world. The remaining hours should be his own. Artificial light has made such a distribution of time possible. The working-periods in many cases may be arranged in the interests of economy, which often means continuous operations. The sun need not be considered when these operations are confined to interiors or localized outdoors.

Thus, artificial light has been an important factor in the great industrial development of the present time. Man now burrows into the earth, navigates under water, travels upon the surface of land and sea, and soars among the clouds piloted by light of his own making. Progress does not halt at sunset but continues twenty-four hours each day. Building, printing, manufacturing, commerce, and other activities are prosecuted continuously, the working-shifts changing at certain periods regardless of the rising or setting sun. Adequate artificial lighting decreases spoilage, increases production, and is a powerful factor in the prevention of industrial accidents.

It has ever been true since the advent of artificial light that the intellect has been largely nourished after the completion of the day's work. The highly developed artificial lighting of the present time may account for much of the vast industry of publication. Books, magazines, and newspapers owe much to convenient and inexpensive artificial light, for without it fewer hours would be available for recreation and advancement through reading. Schools, libraries, and art museums may be attended at night for the betterment of the human race. The immortal Lincoln, it is said, gained his early education largely by the light of the fireplace. But all were not endowed with the persistence of Lincoln, so that illiteracy was more common in his day than in the present age of adequate illumination.

The theatrical stage not only depends for its effectiveness upon artificial light but owes its existence and development largely to this agency. In the moving-picture theater, pictures are projected upon the screen by means of it and even the production of the pictures is independent of daylight. These and a vast number of recreational activities owe much, and in some cases their existence, to artificial light.

Not many centuries ago the streets at night were overrun by thieves and to venture outdoors after dark was to court robbery and even bodily harm. In these days of comparative safety it is difficult to realize the influence that abundant illumination has had in increasing the safety of life and property. Maeterlinck in his poetical drama, "The Bluebird," appropriately has made Light the faithful companion of mankind. The Palace of Night, into which Light is not permitted to enter, is the abode of many evils. Thus the poet has played upon the primitive instincts of the impressiveness of light and darkness.

By combining the symbolism of light, color, and darkness with the instincts which have been inherited by mankind from its superstitious ancestry of the age of mythology, another field of application of artificial light is opened. Light has gradually assumed such attributes as truth, knowledge, progress, enlightenment. Throughout the early ages light was more or less worshiped and thus artificial lights became woven in many religious ceremonies. Some of these have persisted to the present time. The great pageants of peace celebrations and world's expositions appropriately feature artificial light. In drawing upon the potentiality of the expressiveness and impressiveness of light and color, artificial light is playing a major part. Doubtless the future generations will be entertained by gorgeous symphonies of light. Experiments are performed in this direction now and then, and it is reasonable to expect that after many centuries of cultivation of the appreciation of light-symphonies, these will take a place among the arts. The elaborate and complicated music of the present time is appreciated by civilized nations only after many centuries of slow cultivation of taste and understanding.

Light-therapy is to-day a distinct science and art. The germicidal action of light-rays and of some of the invisible rays which ordinarily accompany the luminous rays is well proved. Wounds are treated effectively and water is sterilized by the ultraviolet radiant energy in modern artificial illuminants.

Thousands of lighthouses, light-ships, and light-buoys are scattered along sea-coasts, rivers, and channels. They guide the wheelman and warn the lookout of shoals and reefs. Some of these send forth flashes of light whose intensities are measured in millions of candle-power. Many are unattended for days and even months. These powerful lights dominated by automatic mechanisms have replaced the wood-fires which were maintained a few centuries ago upon certain prominent points.

Signal-lights now guide the railroad train through the night. A burning flare dropped from the rear of a train keeps the following train at a safe distance. Huge search-lights penetrate the night air for many miles. When these are equipped with shutters, a code may be flashed from one ship to another or between the vessel and land. A code from a powerful search-light has been read a hundred miles away because the flashes were projected upon a layer of high clouds and were thus visible far beyond the horizon.

Artificial light played its part in the recent war. Huge search-light equipments were devised for portability. This mobile apparatus was utilized against enemy aircraft and in various other ways. Small hand-lamps are used to send out a pencil of light as directed by a pair of sights and the code is flashed by means of a trigger. Raiding-parties are no longer concealed by the curtain of darkness, for rockets and star-shells are used to illuminate large areas. Flares sent upward to drift slowly downward supported by parachutes saved and cost many lives during the recent war. Rockets are used by ships in distress and also by beleaguered troops.

Experiments are being prosecuted to ascertain the possibilities of artificial light in the forcing of plant-growth, and even chickens are made to work longer hours by its use.

Artificial light is now modified in color or spectral character to meet many requirements. Daylight has been reproduced in spectral quality so that certain processes requiring accurate discrimination of color are now prosecuted twenty-four hours a day under artificial daylight. Colored light is made of the correct quality which does not affect photographic plates of various sensibilities. Monochromatic light is utilized in photo-micrography for the best rendition of detail. Light-waves have been utilized as standards of length because they are invariable and fundamental. Numerous other interesting adaptations of artificial light are in daily use.

This is in reality the age of artificial light, for mankind has not only become independent of daylight in certain respects, but has improved upon natural light. The controllability of artificial light makes it superior to natural light in many ways. In fact, uses have been made of artificial light which are impossible with natural light. Light-sources may be made of a vast variety of shapes, and those may be transported wherever desired. They may be equipped with reflectors and other optical devices to direct or to diffuse the light as required.

Thus, artificial light to-day has numerous advantages over light which has been furnished by the Creator. It is sometimes stated that it can never compete with daylight in cheapness, inasmuch as the latter costs nothing. But this is not true. Even in the residence, daylight costs something, because windows are more expensive than plain walls. The expense of washing windows is an appreciable percentage of the cost of gas or electricity. And there is window-breakage to be considered.

In the more elaborate buildings of the congested portions of cities, daylight is satisfactory a lesser number of hours than in the outlying districts. In some stores, offices, and factories artificial light is used throughout the day. Still, the daylighting-equipment is installed and maintained. Furthermore, when it is considered that much expensive area is given to light-courts and much valuable wall space to windows, it is seen that the cost of daylight in congested cities is in reality considerable. Of course, the daylighting-equipment has value in ventilating, but ventilation may be taken care of in a very satisfactory manner as a separate problem.

The cost of skylights in museums and other large buildings is far greater than that of ordinary ceilings and walls, and the extra allowance for heating is appreciable. The expense of maintenance of some skylights is considerable. Thus it is seen that the cost and maintenance of daylighting-equipment, the loss of valuable rental space and of wall area, and the increased expense of heating are factors which challenge the statement that daylight costs nothing.

In fact, it is not surprising to find that occasionally the elimination of daylighting--the reliance upon artificial light alone--has been seriously contemplated. When the possibilities of the latter are considered, it is reasonable to expect that it will make greater and greater inroads and that many buildings of the future will be equipped solely with artificial-lighting systems.

Naturally, with the tremendous development of artificial light during the present age, a new profession has arisen. The lighting expert is evolving to fill the needs. He is studying the problems of producing and utilizing artificial illumination. He deals with the physics of light-production. His studies of utilization carry him into the vast fields of physiology and psychology. His is a profession which eventually will lead into numerous highways and byways of enterprise, because the possibilities of lighting extend into all those activities which make their appeal to consciousness through the doorway of vision. These possibilities are limited only by the boundaries of human endeavor and in the broadest sense extend even beyond them, for light is one of the most prominent agencies in the scheme of creation. It contributes largely to the safety, the efficiency, and the happiness of civilized beings and beyond all it is a powerful civilizing agency.

II

THE ART OF MAKING FIRE

Scattered over the earth at the present time various stages of civilization are to be found, from the primitive savages to the most highly cultivated peoples. Although it is possible that there are tribes of lowly beings on earth to-day unfamiliar with fire or ignorant of its uses, savages are generally able to make fire. Thus the use of fire may serve the purpose of distinguishing human beings from the lower animals. Surely the savage of to-day who is unable to kindle fire or who possesses a mind as yet insufficiently developed to realize its possibilities, is quite at the mercy of nature's whims. He lives merely by animal prowess and differs little in deeds and needs from the beasts of the jungle. In this imaginary journey to the remote regions beyond the outskirts of civilization it soon becomes evident that the development of artificial light may be a fair measure of civilization.

In viewing the development of artificial light it is seen that preceding the modern electrical age, man depended universally upon burning material. Obviously, the course of civilization has been highly complex and cannot be symbolized adequately by the branching tree. From its obscure beginning far in the impenetrable fog of prehistoric times, it has branched here and there. These various branches have been subjected to many different influences, with the result that some flourished and endured, some retrogressed, some died, some went to seed and fell to take root and to begin again the upward climb. The ultimate result is the varied civilization of the present time, a study of which aids in penetrating the veil that obscures the ages of unrecorded writing. Likewise, material relics of bygone ages supply some threads of the story of human progress and mythology aids in spanning the misty gap between the earliest ages of man and the period when historic writings were begun. Throughout these various stages it becomes manifest that the development of artificial light is associated with the progress of mankind.

According to a certain myth, Prometheus stole fire from heaven and brought this blessing to earth. Throughout the mythologies of various races, fire and, as a consequence, light have been associated with divinity. They have been subjects of worship perhaps more generally than anything else, and these early impressions have survived in the ceremonial uses of light and fire even to the present time. The origin of fire as represented in any of the myths of the superstitious beings of early ages is as suitable as any other, inasmuch as definite knowledge is unavailable. Active volcanoes, spontaneous combustion, friction, accidental focusing of the sun's image, and other means may have introduced primitive beings to fire. A study of savage tribes of the present age combined with a survey of past history of mythology, of material relics, and of the absence of lamps or other lighting utensils leads to the conclusion that the earliest source of light was the wood fire.

Even to-day the savages of remote lands have not advanced further than the wood-fire stage, and they may be found kneeling upon the ground energetically but skilfully rubbing sticks together until the friction kindles a fire. In using these fire-sticks they convert mechanical energy into heat energy. This is a fundamental principle of physics, employed by them as necessity demands, but they are totally ignorant of it as a scientific law. The things which these savages learn are the result of accidental discovery. Until man pondered over such simple facts and coordinated them so that he could extend his knowledge by general reasoning, his progress could not be rapid. But the sluggish mind of primitive man is capable of devising improvements, however slowly, and the art of making fire by means of rubbing fire-sticks gradually became more refined. Mechanical improvements resulted from experience, with the consequence that finally one stick was rubbed to and fro in a groove, or was rapidly twirled between the palms of the hands while one end was pressed firmly into a hole in a piece of wood. In the course of a few seconds or a minute, depending upon skill and other conditions, a fire was obtained. It is interesting to note how civilized man is often compelled by necessity to adopt the methods of primitive beings. The rubbing of sticks is an emergency measure of the master of woodcraft at the present time, and the production of fire in this manner is the proud accomplishment or ambition of every Boy Scout.

Where only such crude means of kindling fire were available it became the custom in some cases to maintain a fire burning continuously in a public place. Around this pyrtaneum the various civil, political, and religious affairs were carried on by the light and warmth of the public fire. Many quaint customs evolved, apparently, from this ancient procedure.

The tinder-box of modern centuries doubtless originated in very early times, for it is inconceivable that the earliest beings did not become aware of the production of sparks when certain stones were struck together. In the stone age, when human beings spent much of their time chiseling implements and utensils from stone by means of tools of the same substance, it appears certain that this means of producing fire was ever apparent. Many of their sharp implements, such as knives and arrow-heads, were made of quartz and similar material and it is likely that the use of two pieces of quartz for producing a spark originated in those remote periods. Alaskan and Aleutian tribes are known to have employed two pieces of quartz covered with native sulphur. When these were struck together with skill, excellent sparks were obtained.

Later, when iron and steel became available, the more modern tinder-box was developed. An early application of the flint-and-steel principle was made by certain Esquimo tribes who obtained fire by striking a piece of quartz against a piece of iron pyrites. The latter is a yellow sulphide of iron, of crystalline form, best known as "fool's gold." Doubtless, the more primitive beings used dried grass, leaves, and moss as inflammable material upon which the sparks were showered. In later centuries the tinder-box was filled with charred grass, linen, and paper. There was a long interval between the development of fire-sticks and that of the tinder-box as measured by the progress of civilization. During recent centuries ordinary brown paper soaked in saltpeter and dried was utilized satisfactorily as an inflammable material. Such devices have been employed in past ages in widely separated regions of the earth. Elaborate specimens of tinder-boxes from Jamaica, Japan, China, Europe, and various other countries are now reposing in the collections in the possession of museums and of individuals.

If the radiant energy from the sun is sufficiently concentrated upon inflammable material, the latter will ignite. Such concentration may be achieved by means of a convex lens or a concave mirror. This method of producing fire does not antedate the more primitive methods such as striking quartz or rubbing wooden sticks, because the materials required are not readily found or prepared, but it is of very remote origin. Aristophanes in his comedy "The Clouds," which is a satire aimed at the science and philosophy of his period (488-385 B. C.), mentions the "burning lens." Nearly every one is familiar with an achievement attributed to Archimedes in which he destroyed the ships at Syracuse by focusing the image of the sun upon them by means of a concave mirror. The ancient Egyptians were proficient in the art of glass-making, so it is likely that the "burning-glass" was employed by them. Even a crude lens of glass will focus an image of the sun sufficiently well to cause inflammable material to ignite.

The energy in sunlight varies enormously, even on clear days, because the water-vapor in the atmosphere absorbs some of the radiant energy emitted by the sun. This absorbed radiation is chiefly known as infra-red energy, which does not arouse the sensation of light. When the water-vapor content of the atmosphere is high, the sun, though it may appear as bright to the eye, in reality is not as hot as it would be if the water-vapor were not present. However, a fire may be kindled by concentrating only the visible rays in sunlight because of the enormous intensity of sunlight. A convex lens fashioned from ice by means of a sharp-edged stone and finally shaped by melting the surfaces as they are rubbed in the palms of the hands, will kindle a fire in highly inflammable material if the sun is high and the atmosphere is fairly clear. Burning-glasses are used to a considerable extent at the present time in certain countries and it is reported that British soldiers were supplied with them during the Boer War. Indicative of the predominant use to which the glass lens was applied in the past is the employment of the term "burning-glass" instead of lens in the scientific writings as late as a century or two ago.

As civilization advanced, leading intellects began to inquire into the mysteries of nature and the periods of pure philosophy gave way to an era of methodical research. Alchemy and superstition began to retire before the attacks of those pioneers who had the temerity to believe that the scheme of creation involved a vast network of invariable laws. In this manner the powerful sciences of physics and chemistry were born a few centuries ago. Among other things the production of fire and light received attention and the "dark ages" were doomed to end. The crude, uncertain, and inconvenient methods of making fire were replaced by steadily improving scientific devices.

Matches were at first cumbersome, dangerous, and expensive, but these gradually evolved into the safety matches of the present time. Although they were primarily intended for lighting fires and various kinds of lamps, billions of them are now used yearly as convenient light-sources. Smoldering hemp or other material treated with niter and other substances was an early form of match used especially for discharging firearms. The modern wax-taper is an evolutionary form of this type of light-source.

Phosphorus has long played a dominant role in the preparation of matches. The first attempt at making them in their modern form appears to have occurred about 1680. Small pieces of phosphorus were used in connection with small splints of wood dipped in sulphur. This type of match did not come into general use until after the beginning of the nineteenth century, owing to its danger and expense. White or yellow phosphorus is a deadly poison; therefore the progress of the phosphorus match was inhibited until the discovery of the relatively harmless form known as red phosphorus. The first commercial application of this form was made in about 1850.

An early ingenious device consisted of a piece of phosphorus contained in a tube. A piston fitted snugly into the tube, by means of which the air could be compressed and the phosphorus ignited. Sulphur matches were ignited from the burning tinder, the latter being fired by flint and steel. In 1828 another form of match consisted of a glass tube containing sulphuric acid and surrounded by a mixture of chlorate of potash and sugar. A pair of nippers was supplied with each box of these "matches," by means of which the tip of the glass tube could be broken off. This liberated the acid, which upon mixing with the other ingredients set fire to them. To this contrivance a roll of paper was attached which was ignited by the burning chemicals.

The lucifer or friction matches appeared in about 1827, but successful phosphorus matches were first made in about 1833. The so-called safety match of the present time was invented in the year 1855. To-day, the total daily output of matches reaches millions and perhaps billions. Automatic machinery is employed in preparing the splints of wood and in dipping them into molten paraffin wax and finally into the igniting composition.

During recent years the principle of the tinder-box has been revived in a device in which sparks are produced by rubbing the mineral cerite (a hydrous silicate of cerium and allied metals) against steel. These sparks ignite a gas-jet or a wick soaked in a highly inflammable liquid such as gasolene or alcohol. This device is a tinder-box of the modern scientific age.

Naturally with the advent of electricity, electrical sparks came into use for lighting gas-jets and mantles and in isolated instances they have served as light-sources. Doubtless, every one is familiar with the parlor stunt of igniting a gas-jet from the discharge from the finger-tips of static electricity accumulated by shuffling the feet across the floor-rug.

Although many of these methods and devices have been used primarily for making fire, they have served as emergency or momentary light-sources. In the outskirts of civilization some of them are employed at the present time and various modern light-sources require a method of ignition.

III

PRIMITIVE LIGHT-SOURCES

Many are familiar with the light of the firefly or of its larv? the glow-worm, but few persons realize that a vast number of insects and lower organisms are endowed with the superhuman ability of producing light by physiological processes. Apparently the chief function of these lighting-plants within the living bodies is not to provide light in the sense that the human being uses it predominantly. That is, these wonderful light-sources seem to be utilized more for signaling, for luring prey, and for protection than for strictly illuminating-purposes. Much study has been given to the production of light by animals, because the secrets will be extremely valuable to mankind. As one floats over tide-water on a balmy evening after dark and watches the pulsating spots of phosphorescent light emitted by the lowly jellyfishes, his imaginative mood formulates the question, "Why are these lowly organisms endowed with such a wonderful ability?"

Despite his highly developed mind and body and his boasted superiority, man must go forth and learn the secrets of light-production before he may emancipate himself from darkness. If man could emit light in relative proportion to his size as compared with the firefly, he would need no other torch in the coal-mine. How independent he would be in extreme darkness where his adapted eyes need only a feeble light-source! Primitive man, desiring a light-source and having no means of making fire, imprisoned the glowing insects in a perforated gourd or receptacle of clay, and thus invented the first lantern perhaps before he knew how to make fire. The fireflies of the West Indies emit a continuous glow of considerable luminous intensity and the natives have used these imprisoned insects as light-sources. Thus mankind has exhibited his superiority by adapting the facilities at hand to the growing requirements which his independent nature continuously nourished. His insistent demand for independence in turn has nourished his desire to learn nature's secrets and this desire has increased in intensity throughout the ages.

The act of imprisoning a glowing insect was in itself no greater stride along the highway of progress than the act of picking a tasty fruit from its tree. However, the crude lantern perhaps directed his primitive mind to the possibilities of artificial light. The flaming fagot from the fire was the ancestor of the oil-lamp, the candle, the lantern, and the electric flash-light. It is a matter of conjecture how much time elapsed before his feeble intellect became aware that resinous wood afforded a better light-source than woods which were less inflammable. Nevertheless, pine knots and similar resinous pieces of wood eventually were favored as torches and their use has persisted until the present time. In some instances in ancient times resin was extracted from wood and burned in vessels. This was the forerunner of the grease-and the oil-lamp. In the woods to-day the craftsman of the wilds keeps on the lookout for live trees saturated with highly inflammable ingredients.

Viewed from the present age, these smoking, flickering light-sources appear very crude; nevertheless they represent a wide gulf between their users and those primitive beings who were unacquainted with the art of making fire. Although the wood fire prevailed as a light-source throughout uncounted centuries, it was subjected to more or less improvement as civilization advanced. When the wood fire was brought indoors the day was extended and early man began to develop his crude arts. He thought and planned in the comfort and security of his cave or hut. By the firelight he devised implements and even decorated his stone surroundings with pictures which to-day reveal something of the thoughts and activities of mankind during a civilization which existed many thousand years ago.

When it was too warm to have a roaring fire upon the hearth, man devised other means for obtaining light without undue warmth. He placed glowing embers upon ledges in the walls, upon stone slabs, or even upon suspended devices of non-inflammable material. Later he split long splinters of wood from pieces selected for their straightness of grain. These burning splinters emitting a smoking, feeble light were crude but they were refinements of considerable merit. A testimonial of their satisfactoriness is their use throughout many centuries. Until very recent times the burning splinter has been in use in Scotland and in other countries, and it is probable that at present in remote districts of highly civilized countries this crude device serves the meager needs of those whose requirements have been undisturbed by the progress of civilization. Scott, in "The Legend of Montrose," describes a table scene during a feast. Behind each seat a giant Highlander stood, holding a blazing torch of bog-pine. This was also the method of lighting in the Homeric age.

Crude clay relics representing a human head, from the mouth of which the wood-splinters projected, appear to corroborate the report that the flaming splinter was sometimes held in the mouth in order that both hands of a workman would be free. Splinter-holders of many types have survived, but most of them are of the form of a crude pedestal with a notch or spring clip at its upper end. The splinter was held in this clip and burned for a time depending upon its length and the character of the wood. It was the business of certain individuals to prepare bundles of splinters, which in the later stages of civilization were sold at the market-place or from house to house. Those who have observed the frontiersman even among civilized races will be quite certain that the wood for splinters was selected and split with skill, and that the splinters were burned under conditions which would yield the most satisfactory light. It is a characteristic of those who live close to nature, and are thus limited in facilities, to acquire a surprising efficiency in their primitive activities.

An obvious step in the use of burning wood as a light-source was to place such a fire on a shelf or in a cavity in the wall. Later when metal was available, gratings or baskets were suspended from the ceiling or from brackets and glowing embers or flaming chips were placed upon them. Some of these were equipped with crude chimneys to carry away the smoke, and perhaps to increase the draft. In more recent centuries the first attempt at lighting outdoor public places was by means of metal baskets in which flaming wood emitted light. It was the duty of the watchman to keep these baskets supplied with pine knots. In early centuries street-lighting was not attempted, and no serious efforts worthy of consideration as adequate lighting were made earlier than about a century ago. As a consequence the "link-boy" came into existence. With flaming torch he would escort pedestrians to their homes on dark nights. This practice was in vogue so recently that the "link-boy" is remembered by persons still living. In England the profession appears to have existed until about 1840.

Somewhat akin to the wood-splinter, and a forerunner of the candle, was the rushlight. In burning wood man noticed that a resinous or fatty material increased the inflammability and added greatly to the amount of light emitted. It was a logical step to try to reproduce this condition by artificial means. As a consequence rushes were cut and soaked in water. They were then peeled, leaving lengths of pith partially supported by threads of the skin which were not stripped off. These sticks of pith were placed in the sun to bleach and to dry, and after they were thoroughly dry they were dipped in scalding grease, which was saved from cooking operations or was otherwise acquired for the purpose. A reed two or three feet long held in the splinter-holder would burn for about an hour. Thus it is seen that man was beginning to progress in the development of artificial light. In developing the rushlight he was laying the foundation for the invention of the candle. Pliny has mentioned the burning of reeds soaked in oil as a feature of funeral rites. Many crude forerunners of the candle were developed in various parts of the world by different races. For example, the Malays made a torch by wrapping resinous gum in palm leaves, thus devising a crude candle with the wick on the outside.

Many primitive uses of vegetable and animal fats were forerunners of the oil-lamp. In the East Indies the candleberry, which contains oily seeds, has been burned for light by the natives. In many cases burning fish and birds have served as lamps. In the Orkney Islands the carcass of a stormy petrel with a wick in its mouth has been utilized as a light-source, and in Alaska a fish in a split stick has provided a crude torch for the natives. These primitive methods of obtaining artificial light have been employed for centuries and many are in use at the present time among uncivilized tribes and even by civilized beings in the remote outskirts of civilization. Surely progress is limited where a burning fish serves as a torch, or where, at best, the light-sources are feeble, smoking, flickering, and ill-smelling!

Progress insisted upon a light-source which was free from the defects of the crude devices already described and the next developments were improvements to the extent at least that combustion was more thorough. The early oil-lamps and candles did not emit much smoke, but they were still feeble light-sources and not always without noticeable odors. Nevertheless, they marked a tremendous advance in the production of artificial light. Although they were not scientific developments in the modern sense, the early oil-lamp and the candle represented the great possibilities of utilizing knowledge rather than depending upon the raw products of nature in unmodified forms. The advent of these two light-sources in reality marked the beginning of the civilization which was destined to progress and survive.

Although such primitive light-sources as the flaming splinter and the glowing ember have survived until the present age, lamps consisting of a wick dipped into a receptacle containing animal and vegetable oils have been in use among the more advanced peoples since prehistoric times. Oil-lamps are to be seen in the earliest Roman illustrations. During the height of ancient civilization along the eastern shores of the Mediterranean Sea, elaborate lighting was effected by means of the shallow grease-or oil-lamp. It is difficult to estimate the age in which this form of light-source originated, but some lamps in existence in collections at the present time appear to have been made as early as four or five thousand years before the Christian era. It is noteworthy that such lamps did not differ materially in essential details from those in use as late as a few centuries ago.

At first the grease used was the crude fat from animals. Vegetable oils also were burned in the early lamps. The Japanese, for example, extracted oil from nuts. When the demands of civilization increased, extensive efforts were made to obtain the required fats and oils. Amphibious animals of the North and the huge mammals of the sea were slaughtered for their fat, and vegetable sources were cultivated. Later, sperm and colza were the most common oils used by the advanced races. The former is an animal oil obtained from the head cavities of the sperm-whale; the latter is a vegetable oil obtained from rape-seed. Mineral oil was introduced as an illuminant in 1853, and the modern lamp came into use.

The grease-and oil-lamps in general were of such a form that they could be carried with ease and they had flat bottoms so that they would rest securely. The simplest forms had a single wick, but in others many wicks dipped into the same receptacle. The early ones were of stone, but later, lamps were modeled from clay or terra cotta and finally from metals. They were usually covered and the wick projected through a hole in the top near the edge. Large stone vases filled with a hundred pounds of liquid fat are known to have been used in early times. As a part of the setting in the celebration of festivals the ancient nations of Asia and Africa placed along the streets bronze vases filled with liquid fat. The Esquimaux to-day use this form of lamp, in which whale-oil and seal blubber is the fuel. Incidentally, these lamps also supply the only artificial heat for their huts and igloos. The heat from these feeble light-sources and from their bodies keeps these natives of the arctics warm within the icy walls of their abodes.

Very beautiful oil-lamps of brass, bronze, and pewter evolved in such countries as Egypt. Many of these were designed for and used in religious ceremonies. The oil-lamps of China, Scotland, and other countries in later centuries were improved by the addition of a pan beneath the oil-receptacle, to catch drippings from the wick or oil which might run over during the filling.

The Chinese lamps were sometimes made of bamboo, but the Scottish lamps were made of metal. A flat metal lamp, called a crusie, was one of the chief products of blacksmiths and was common in Scotland until the middle of the nineteenth century. This type of lamp was used by many nations and has been found in the catacombs of Rome. The crusie was usually suspended by an iron hook and the flow of oil to the wick could be regulated by tilting. The wick in the Scottish lamps consisted of the pith of rushes, cloth, or twisted threads. These early oil-lamps were almost always shallow vessels into which a short wick was dipped, and it was not until the latter part of the eighteenth century that other forms came into general use. The change in form was due chiefly to the introduction of scientific knowledge when mineral oil was introduced. As early as 1781 the burning of naptha obtained by distilling coal at low temperatures was first discussed, but no general applications were made until a later period. This was the beginning of many marked improvements in oil-lamps, and was in reality the birth of the modern science of light-production.

As the activities of man became more complex he met from his growing store of knowledge the increasing requirements of lighting. In consequence, many ingenious devices for lighting were evolved. For example, in England in the seventeenth century man was already burrowing into the earth for coal and of course encountered coal-gases. These inflammable gases were first known for the direful effects which they so often produced rather than for their useful qualities. Although they were known to miners long before they received scientific attention, the earliest account of them in the Transactions of the Royal Society was presented in the year 1667. A description of early gas-lighting has been reserved for a later chapter, but the foregoing is noted at this point to introduce a novel early method of lighting in coal-mines where inflammable gases were encountered. In discussing this coal-gas another early writer stated that "it will not take fire except by flame" and that "sparks do not affect it." One of the early solutions of the problem of artificial lighting under such conditions is summarized as follows:

Before the invention of Sir Humphrey Davy's Safety Lamp, this property of the gas gave rise to a variety of contrivances for affording the miners sufficient light to pursue their operations; and one of the most useful of these inventions was a mill for producing light by sparks elicited by the collision of flint and steel.

Such a stream of sparks may appear a very crude and unsatisfactory solution as judged by present standards, but it was at least an ingenious application of the facilities available at that time. Various other devices were resorted to in the coal-mines before the introduction of a safety lamp.

In discussing the candle it is necessary again to go back to an early period, for it slowly evolved in the course of many centuries. It is the natural descendant of the rushlight, the grease-lamp, and various primitive devices. Until the advent of the more scientific age of artificial lighting, the candle stood preeminent among early light-sources. It did not emit appreciable smoke or odor and it was conveniently portable and less fragile than the oil-lamp. Candles have been used throughout the Christian era and some authorities are inclined to attribute their origin to the Phoenicians. It is known that the Romans used them, especially the wax-candles, in religious ceremonies. The Phoenicians introduced them into Byzantium, but they disappeared under the Turkish rule and did not come into use again until the twelfth century.

The wax-candle was very much more expensive than the tallow-candle until the fifteenth century, when its relative cost was somewhat reduced, bringing it within the means of a greater proportion of the people. Nevertheless it has long been used, chiefly by the wealthy; the departing guest of the early Victorian inn would be likely to find an item on his bill such as this: "For a gentleman who called himself a gentleman, wax-lights, 5/." Poor men used tallow dips or went to bed in the dark. It is interesting to note the importance of the candle in the household budget of early times in various sayings. For example, "The game is not worth the candle," implies that the cost of candle-light was not ignored. In these days little attention is given to the cost of artificial light under similar conditions. If a person "burns a candle at both ends" he is wasteful and oblivious to the consequences of extravagance whether in material goods or in human energy.

With the rise of the Christian church, candles came to be used in religious ceremonies and many of the symbolisms, meanings, and customs survive to the present time. Some of the finest art of past centuries is found in the old candlesticks. Many of these antiques, which ofttimes were gifts to the church, have been preserved to posterity by the church. The influence of these lighting accessories is often noted in modern lighting-fixtures, but unfortunately early art often suffers from adaptation to the requirements of modern light-sources, or the eyesight suffers from a senseless devotion to art which results in the use of modern light-sources, unshaded and glaring, in places where it was unnecessary to shade the feeble candle.

The oldest materials employed for making candles are beeswax and tallow. The beeswax was bleached before use. The tallow was melted and strained and then cotton or flax fibers were dipped into it repeatedly, until the desired thickness was obtained. In early centuries the pith of rushes was used for wicks. Tallow is now used only as a source of stearine. Spermaceti, a fatty substance obtained from the sperm-whale, was introduced into candle-making in about 1750 and great numbers of men searched the sea to fill the growing demands. Paraffin wax, a mixture of solid hydrocarbons obtained from petroleum, came into use in 1854 and stearine is now used with it. The latter increases the rigidity and decreases the brittleness of the candle. Some of the modern candles are made of a mixture of stearine and the hard fat extracted from cocoanut-oil. Modern candles vary in composition, but all are the product of much experience and of the application of scientific knowledge. The wicks are now made chiefly of cotton yarn, braided or plaited by machinery and chemically treated to aid in complete combustion when the candle is burned. Their structure is the result of long experience and they are now made so that they bend and dip into the molten fuel and are wholly consumed. This eliminates the necessity of trimming.

Candles have been made in various ways, including dipping, pouring, drawing, and molding. Wax-candles are made by pouring, because wax cannot be molded satisfactorily. Drawing is somewhat similar to dipping, except that the process is more or less continuous and is carried out by machinery. Molding, as the term implies, involves the use of molds, of the size and shape desired.

The candlestick evolved from the most primitive wooden objects to elaborately designed and decorated works of art. The primitive candlestick was crude and was no more than a holder of some kind for keeping the candle upright. Later a form of cup was attached to the stem of the holder, to catch the dripping wax or fat. The latter improvement has persisted throughout the centuries. The modern candle is by no means an unsatisfactory light-source. Those who have had experience with it in the outskirts of civilization will testify that it possesses several desirable characteristics. Supplies of candles are transported without difficulty; the lighted candle is easily carried about; and the light in a quiescent atmosphere is quite satisfactory, if common sense is used in shading and placing the candle. Although in a sense a primitive light-source, it is a blessing in many cases and, incidentally, it is extensively used to-day in industries, in religious ceremonies, as a decorative element at banquets, and in the outposts of civilization.

This account of the evolution of light-sources has crossed the threshold of what may be termed modern scientific light-production in the case of the candle and the oil-lamp. There is a period of a century or more during which scientific progress was slow, but those years paved the way for the extraordinary developments of the last few decades.

IV

THE CEREMONIAL USE OF LIGHT

Inasmuch as the symbolisms and ceremonial uses of light originated in the childhood of the human race and were nourished throughout the age of mythology, the early light-sources are associated more with this phase of artificial light than modern ones. For this reason it appears appropriate to present this discussion before entering into the later stages of the development and utilization of artificial light. Furthermore, many of the traditions of lighting at the present time are survivors of the early ages. Lighting-fixtures show the influence of this byway of lighting, and in those cases where the ceremonial use of light has survived to the present time, modern light-sources cannot be employed wisely in replacing more primitive ones without consideration of the origin and existence of the customs. In fact, candles are likely to be used for hundreds of years to come, owing to the sentiment connected with them and to the established customs founded upon centuries of traditional use.

Doubtless, the sun as a source of heat and light and of the blessings which these bring to earth, is responsible largely for the divine significance bestowed upon light. Darkness very deservingly acquired many uncomplimentary attributes, for danger lurked behind its veil and it was the suitable abode of evil spirits. It harbored all that was the antithesis of goodness, happiness, and security. Light naturally became sacred, life-giving, and symbolic of divine presence. Fire was to primitive beings the most impressive phenomenon over which they had any control, and it was sufficiently mysterious in its operation to warrant a connection with the supernatural. Thus it was very natural that these earlier beings worshiped it as representing divine presence. The sun, as Ra, was one of the chief gods of the ancient Egyptians; and the Assyrians, the Babylonians, the ancient Greeks, and many other early peoples gave a high place to this deity. Among simpler races the sun was often the sole object of worship, and those peoples who worship Light as the god of all, in a sense are not far afield. Fire-worshipers generally considered fire as the purest representation of heavenly fire, the origin of everything that lives.

Light was considered such a blessing that lamps were buried with the dead in order that spirits should be able to have it in the next world. This custom has prevailed widely but the fact that the lamps were unlighted indicates that only the material aspect was considered. It is interesting to note that the lamps and other light-sources in pagan temples and religious processions were not symbolical but were offerings to the gods. In later centuries a deeper symbolical meaning became attached to light and burning lamps were placed upon the tombs of important personages. The burying of lamps with the dead appears to have originated in Asia. The Phoenicians and Romans apparently continued the custom, but no traces of it have been found in Greece and Egypt.

Fire and light have been closely associated in various religious creeds and their ceremonies. The Hindu festival in honor of the goddess of prosperity is attended by the burning of many lamps in the temples and homes. The Jewish synagogues have their eternal lamps and in their rituals fire and light have played prominent roles. The devout Brahman maintains a fire on the hearth and worships it as omniscient and divine. He expects a brand from this to be used to light his funeral pyre, whose fire and light will make his spirit fit to enter his heavenly abode. He keeps a fire burning on the altar, worships Agni, the god of fire, and makes fire sacrifices on various occasions such as betrothals and marriages. To the Mohammedans lighted lamps symbolize holy places, and the Kaaba at Mecca, which contains a black stone supposed to have been brought from heaven, is illuminated by thousands of lamps. Many of the uses to which light was put in ancient times indicate its rarity and sacred nature. Doubtless, the increasing use of artificial light at festivals and celebrations of the present time is partly the result of lingering customs of bygone centuries and partly due to a recognition of an innate appeal or attribute of light. Certainly nothing is more generally appropriate in representing joy and prosperity.

Throughout all countries ancient races had woven natural light and fire into their rites and customs, so it became a natural step to utilize artificial light and fire in the same manner. It would be tedious and monotonous to survey the vast field of ancient worship of light, for the underlying ideas are generally similar. The mythology of the Greeks is illustrative of the importance attached to fire and light by the cultivated peoples of ancient times. The myth of Prometheus emphasizes the fact that in those remote periods fire and light were regarded as of prime importance. According to this myth, fire and light were contained in heaven and great cunning and daring were necessary in order to obtain it. Prometheus stole this heavenly fire, for which act he was chained to the mountain and made to suffer. The Greeks mark this event as the beginning of human civilization. All arts are traced to Prometheus, and all earthly woe likewise. As past history is surveyed it appears natural to think of scientific men who have become martyrs to the quest of hidden secrets. They have made great sacrifices for the future benefit of civilization and not a few of them have endured persecution even in recent times. The Greeks recognized that a new era began with the acquisition of artificial light. Its divine nature was recognized and it became a phenomenon for worship and a means for representing divine presence. The origin of fire and light made them holy. The fire on the altar took its place in religious rites and there evolved many ceremonial uses of lamps, candles, and fire.

The Greeks and Romans burned sacred lamps in the temples and utilized light and fire in many ceremonies. The torch-race, in which young men ran with lighted torches, the winner being the one who reached the goal first with his torch still alight, originated in a Grecian ceremony of lighting the sacred fire. There are many references in ancient Roman and Grecian literature to sacred lamps burning day and night in sanctuaries and before statues of gods and heroes. On birthdays and festivals the houses of the

Romans were specially ornamented with burning lamps. The Vestal Virgins in Rome maintained the sacred fire which had been brought by fugitives from Troy. In ancient Rome when the fire in the Temple of Vesta became extinguished, it was rekindled by the rubbing of a piece of wood upon another until fire was obtained. This was carried into the temple by the Vestal Virgin and the sacred fire was rekindled. The fire produced in this manner, for some reason, was considered holy.

The early peoples displayed many lamps on feast-days and an example of extravagance in this respect is an occasion when King Constantine commanded that the entire city of Constantinople be illuminated by wax-candles on Christmas Eve. Candelabra, of the form of the branching tree, were commonly in use in the Roman temples.

The ceremonial use of light in the Christian church evolved both from adaptations of pagan customs and of the natural symbolisms of fire and light. However, these acquired a deeper meaning in Christianity than in early times because they were primarily visible representations or manifestations of the divine presence. The Bible contains many references to the importance and symbolisms of light and fire. According to the First Book of Moses, the achievement of the Creator immediately following the creation of "the heavens and the earth" was the creation of light. The word "light" is the forty-sixth word in Genesis. Christ is "the true light" and Christians are "children of light" in war against the evil "powers of darkness." When St. Paul was converted "there shined about him a great light from heaven." The impressiveness and symbolism of fire and light are testified to in many biblical expressions. Christ stands "in the midst of seven candle-sticks" with "his eyes as a flame of fire." When the Holy Ghost appeared before the apostles "there appeared unto them cloven tongues of fire." When St. Paul was preaching the gospel of Christ at Alexandria "there were many lights" suggesting a festive illumination.

According to the Bible, the perpetual fire which came originally from heaven was to be kept burning on the altar. It was holy and those whose duty it was to keep it burning were guilty of a grave offense if they allowed it to be extinguished. If human hands were permitted to kindle it, punishment was meted out. The two sons of Aaron who "offered strange fire before the Lord" were devoured by "fire from the Lord." The seven-branched candlestick was lighted eternally and these burning light-sources were necessary accompaniments of worship.

The countless ceremonial uses of fire and light which had evolved in the past centuries were bound to influence the rites and customs of the Christian church. The festive illumination of pagan temples in honor of gods was carried over into the Christian era. The Christmas tree of to-day is incomplete without its many lights. Its illumination is a homage of light to the source of light. The celebration of Easter in the Church of the Holy Sepulchre in Jerusalem is a typical example of fire-worship retained from ancient times. At the climax of the services comes the descent of the Holy Fire. The central candelabra suddenly becomes ablaze and the worshipers, each of whom carries a wax taper, light their candles therefrom and rush through the streets. The fire is considered to be of divine origin and is a symbol of resurrection. The custom is similar in meaning to the light which in older times was maintained before gods.

During the first two or three centuries of the Christian era the ceremonial use of light does not appear to have been very extensive. Writings of the period contain statements which appear to ridicule this use to some extent. For example, one writer of the second century states that "On days of rejoicing ... we do not encroach upon daylight with lamps." Another, in the fourth century, refers with sarcasm to the "heathen practice" in this manner: "They kindle lights as though to one who is in darkness. Can he be thought sane who offers the light of lamps and candles to the Author and Giver of all light?"

That candles were lighted in cemeteries is evidenced by an edict which forbade their use during the day. Lamps of the early centuries of the Christian era have been found in the catacombs of Rome which are thought to have been ceremonial lamps, for they were not buried with the dead. They were found only in niches in the walls. During these same centuries elaborate candelabra containing hundreds of candles were kept burning before the tombs of saints. Notwithstanding the doubt that exists as to the extent of ceremonial lighting in the early centuries of the Christian era, it is certain that by the beginning of the fifth century the ceremonial use of light in the Christian church had become very extensive and firmly established. That this is true and that there were still some objections is indicated by many controversies. Some thought that lamps before tombs were ensigns of idolatry and others felt that no harm was done if religious people thus tried to honor martyrs and saints. Some early writings convey the idea that the ritualistic use of lights in the church arose from the retention of lights necessary at nocturnal services after the hours of worship had been changed to daytime.

Passing beyond the early controversial period, the ceremonial use of light is everywhere in evidence at ordinary church services. On special occasions such as funerals, baptisms, and marriages, elaborate altar-lighting was customary. The gorgeous candelabra and the eternal lamp are noted in many writings. Early in the fifth century the pope ordered that candles be blessed and provided rituals for this ceremony. Shortly after this the Feast of Purification of the Virgin was inaugurated and it became known as Candlemas because on this day the candles for the entire year were blessed. However, it appears that the blessing of candles was not carried out in all churches. Altar lights were not generally used until the thirteenth century. They were originally the seven candles carried by church officials and placed near the altar.

The custom of placing lighted lamps before the tombs of martyrs was gradually extended to the placing of such lamps before various objects of a sacred or divine relation. Finally certain light-sources themselves became objects of worship and were surrounded by other lamps, and the symbolisms of light grew apace. A bishop in the sixth century heralded the triple offering to God represented by the burning wax-candle. He pointed out that the rush-wick developed from pure water; that the wax was the product of virgin bees; and that the flame was sent from heaven. Each of these, he was certain, was an offering acceptable to God. Wax-candles became associated chiefly with religious ceremonies. The wax later became symbolic of the Blessed Virgin and of the body of Christ. The wick was symbolical of Christ's soul, the flame represented his divine character, and the burning candle thus became symbolical of his death. The lamp, lantern, and taper are frequently symbols of piety, heavenly wisdom, or spiritual light. Fire and flames are emblems of zeal and fervor or of the sufferings of martyrdom and the flaming heart symbolizes fervent piety and spiritual or divine love.

By the time the Middle Ages were reached the ceremonial uses of light became very complex, but for the Roman Catholic Church they may be divided into three general groups: (1) They were symbolical of God's presence or of the effect of his presence; of Christ or of "the children of light"; or of joy and content at festivals. (2) They may be offered in fulfillment of a religious vow; that is, as an act of worship. (3) They may possess certain divine power because of their being blessed by the church, and therefore may be helpful to soul and body. The three conceptions are indicated in the prayers offered at the blessing of the candles on Candlemas as follows: (1) "O holy Lord ... who ... by thy command didst cause this liquid to come by the labor of bees to the perfection of wax, ... we beseech thee ... to bless and sanctify these candles for the use of men, and the health of bodies and souls...." (2) "...these candles, which we thy servants desire to carry lighted to magnify thy name; that by offering them to thee, being worthily inflamed with the holy fire of thy most sweet charity, we may deserve...." (3) "O Lord Jesus Christ, the true light, ... mercifully grant, that as these lights enkindled with visible fire dispel nocturnal darkness, so our hearts illuminated by visible fire," etc.

In general, the ceremonial uses of lights in this church were originated as a forceful representation of Christ and of salvation. On the eve of Easter a new fire, emblematic of the arisen Christ, is kindled, and all candles throughout the year are lighted from this. During the service of Holy Week thirteen lighted candles are placed before the altar and as the penitential songs are sung they are extinguished one by one. When but one remains burning it is carried behind the altar, thus symbolizing the last days of Christ on earth. It is said that this ceremony has been traced to the eighth century. On Easter Eve, after the new fire is lighted and blessed, certain ceremonies of light symbolize the resurrection of Christ. From this new fire three candles are lighted and from these the Paschal Candle. The origin of the latter is uncertain, but it symbolizes a victorious Christ. From it all the ceremonial lights of the church are lighted and they thereby are emblematic of the presence of the light of Christ.

Many interesting ceremonial uses may be traced out, but space permits a glimpse of only a few. At baptismal services the paschal candle is dipped into the water so that the latter will be effective as a regenerative element. The baptized child is reborn as a child of light. Lighted candles are placed in the hands of the baptized persons or of their god-parents. Those about to take vows carry lights before the church official and the same idea is attached to the custom of carrying or of holding lights on other occasions such as weddings and first communion. Lights are placed around the bodies of the dead and are carried at the funeral. They not only protect the dead from the powers of darkness but they symbolize the dead as still living in the light of Christ. The use of lighted candles around bodies of the dead still survives to some extent among Protestants, but their significance has been lost sight of. Even in the eighteenth century funerals in England were accompanied by lighted tapers, but the carrying of lights in other processions appears to have ceased with the Reformation. In some parts of Scotland it is still the custom to place two lighted candles on a table beside a corpse on the day of the funeral.

With the importance of light in the ritual of the church it is not surprising that the extinction of lights is a part of the ceremony of excommunication. Such a ceremony is described in an early writing thus: "Twelve priests should stand about the bishop, holding in their hands lighted torches, which at the conclusion of the anathema or excommunication they should cast down and trample under foot." When the excommunicant is reinstated, a lighted candle is placed in his hands as a symbol of reconciliation. These and many other ceremonial uses of light have been and are practised, but they are not always mandatory. Furthermore, the customs have varied from time to time, but the few which have been touched upon illustrate the impressive part that light has played in religious services.

During the Reformation the ceremonial use of lights was greatly altered and was abolished in the Protestant churches as a relic of superstition and papal authority. In the Lutheran churches ceremonial lights were largely retained, in the Church of England they have been subjected to many changes largely through the edicts of the rulers. In the latter church many controversies were waged over ceremonial lights and their use has been among the indictments of a number of officials of the church in impeachment cases before the House of Commons. Many uses of light in religious ceremonies were revived in cathedrals after the Restoration and they became wide-spread in England in the nineteenth century. As late as 1889 the Archbishop of Canterbury ruled that certain ceremonial candles were lawful according to the Prayer-Book of Edward VI, but the whole question was left open and unsettled.

These byways of artificial light are complex and fascinating because their study leads into many channels and far into the obscurity of the childhood of the human race. A glimpse of them is important in a survey of the influence of artificial light upon the progress of civilization because in these usages the innate and acquired impressiveness of light is encountered. Although many ceremonial uses of light remain, it is doubtful if their significance and especially their origin are appreciated by most persons. Nevertheless, no more interesting phase of artificial light is encountered than this, which reaches to the foundation of civilization.

V

OIL-LAMPS OF THE NINETEENTH CENTURY

It will be noted that the light-sources throughout the early ages were flames, the result of burning material. This principle of light-production has persisted until the present time, but in the latter part of the nineteenth century certain departures revolutionized artificial lighting. However, it is not the intention to enter the modern period in this chapter except in following the progress of the oil-lamp through its period of scientific development. The oil-lamp and the candle were the mainstays of artificial lighting throughout many centuries. The fats and waxes which these light-sources burned were many but in the later centuries they were chiefly tallow, sperm-oil, spermaceti, lard-oil, olive-oil, colza-oil, bees-wax and vegetable waxes. Those fuels which are not liquid are melted to liquid form by the heat of the flame before they are actually consumed. The candle is of the latter type and despite its present lowly place and its primitive character, it is really an ingenious device. Its fuel remains conveniently solid so that it is readily shipped and stored; there is nothing to spill or to break beyond easy repair; but when it is lighted the heat of its flame melts the solid fuel and thus it becomes an "oil-lamp." Animal and vegetable oils were mainly used until the middle of the nineteenth century, when petroleum was produced in sufficient quantities to introduce mineral oils. This marked the beginning of an era of developments in oil-lamps, but these were generally the natural offspring of early developments by Ami Argand.

Before man discovered that nature had stored a tremendous supply of mineral oil in the earth he was obliged to hunt broadcast for fats and waxes to supply him with artificial light. He also was obliged to endure unpleasant odors from the crude fuels and in early experiments with fats and waxes the odor was carefully noted as an important factor. Tallow was a by-product of the kitchen or of the butcher. Stearine, a constituent of tallow, is a compound of glyceryl and stearic acid. It is obtained by breaking up chemically the glycerides of animal fats and separating the fatty acids from glycerin. Fats are glycerides; that is, combinations of oleic, palmetic, and stearic acids. Inasmuch as the former is liquid at ordinary temperatures and the others are solid, it follows that the consistency or solidity of fats depend upon the relative proportions of the three constituents. The sperm-whale, which lives in the warmer parts of all the oceans, has been hunted relentlessly for fuels for artificial lighting. In its head cavities sperm-oil in liquid form is found with the white waxy substance known as spermaceti. Colza-oil is yielded by rape-seed and olive-oil is extracted from ripe olives. The waxes are combinations of allied acids with bases somewhat related to glycerin but of complex composition. Fats and waxes are more or less related, but to distinguish them carefully would lead far afield into the complexities of organic chemistry. All these animal and vegetable products which were used as fuels for light-sources are rich in carbon, which accounts for the light-value of their flames. The brightness of such a flame is due to incandescent carbon particles, but this phase of light-production is discussed in another chapter. These oils, fats, and waxes are composed by weight of about 75 to 80 per cent. carbon; 10 to 15 per cent. hydrogen; and 5 to 10 per cent. oxygen.

Until the middle of the eighteenth century the oil-lamps were shallow vessels filled with animal or vegetable oil and from these reservoirs short wicks projected. The flame was feeble and smoky and the odors were sometimes very repugnant. Viewing such light-sources from the present age in which light is plentiful, convenient, and free from the great disadvantages of these early oil-lamps, it is difficult to imagine the possibility of the present civilization emerging from that period without being accompanied by progress in light-production. The improvements made in the eighteenth century paved the way for greater progress in the following century. This is the case throughout the ages, but there are special reasons for the tremendous impetus which light-production has experienced in the past half-century. These are the acquirement of scientific knowledge from systematic research and the application of this knowledge by organized development.

The first and most notable improvement in the oil-lamp was made by

Argand in 1784. Our nation was just organizing after its successful struggle for independence at the time when the production of light as a science was born. Argand produced the tubular wick and contributed the greatest improvement by being the first to perform the apparently simple act of placing a glass chimney upon the lamp. His burner consisted of two concentric metal tubes between which the wick was located. The inner tube was open, so that air could reach the inner surface of the wick as well as the outer surface. The lamp chimney not only protected the flame from drafts but also improved combustion by increasing the supply of air. It rested upon a perforated flange below the burner. If the glass chimney of a modern kerosene lamp be lifted, it will be noted that the flame flickers and smokes and that it becomes steady and smokeless when the chimney is replaced. The advantages of such a chimney are obvious now, but Argand for his achievements is entitled to a place among the great men who have borne the torch of civilization. He took the first step toward adequate artificial light and opened a new era in lighting.

The various improvements of the oil-lamp achieved by Argand combined to effect complete combustion, with the result that a steady, smokeless lamp of considerable luminous intensity was for the first time available. Many developments followed, among which was a combination of reservoir and gravity feed which maintained the oil at a constant level. In later lamps, upon the adoption of mineral oil, this was found unnecessary, perhaps owing to the construction of the wick and to the physical characteristics of the oil which favored capillary action in the wick. However, the height of the oil in the reservoir of modern oil-lamps makes some difference in the amount of light emitted.

The Carcel lamp, which appeared in 1800, consisted of a double piston operated by clockwork. This forced the oil through a tube to the burner. Franchot invented the moderator lamp in 1836, which, because of its simplicity and efficiency soon superseded many other lamps designed for burning animal and vegetable oils. The chief feature of the moderator lamp is a spiral spring which forces the oil upward through a vertical tube to the burner. These are still used to some extent in France, but owing to the fact that "mechanical" lamps eventually were very generally replaced by more simple ones, it does not appear necessary to describe these complex mechanisms in detail.

When coal is distilled at moderate temperatures, volatile liquids are obtained. These hydrocarbons, being inflammable, naturally attracted attention when first known, and in 1781 their use as fuel for lamps was suggested. However, it was not until 1820 that the light oils obtained by distilling coal-tar, a by-product of the coal-gas industry which was then in its early stage of development, were burned to some extent in the Holliday lamp. In this lamp the oil is contained in a reservoir from the bottom of which a fine metal tube carries the oil down to a rose-burner. The oil is heated by the flame and the vaporized mineral oil which escapes through small orifices is burned. This type of lamp has undergone many physical changes, but its principle survives to the present time in the gasolene and kerosene burners hanging on a pole by the side of the street-peddler's stand.

Although petroleum products were not used to any appreciable extent for illuminating-purposes until after the middle of the nineteenth century, mineral oil is mentioned by Herodotus and other early writers. In 1847 petroleum was discovered in a coal-mine in England, but the supply failed in a short time. However, the discoverer, James Young, had found that this oil was valuable as a lubricant and upon the failure of this source he began experiments in distilling oil from shale found in coal deposits. These were destined to form the corner-stone of the oil industry in Scotland. In 1850 he began producing petroleum in this manner, but it was not seriously considered for illuminating-purposes. However, in Germany about this time lamps were developed for burning the lighter distillates and these were introduced into several countries. But the price of these lighter oils was so great that little progress was made until, in 1859, Col. E. L. Drake discovered oil in Pennsylvania. By studying the geological formations and concluding that oil should be obtained by boring, Drake gave to the world a means of obtaining petroleum, and in quantities which were destined to reduce the price of mineral oil to a level undreamed of theretofore. To his imagination, which saw vast reservoirs of oil in the depths of the earth, the world owes a great debt. Lamps were imported from Germany to all parts of the civilized world and the kerosene lamp became the prevailing light-source. Hundreds of American patents were allowed for oil-lamps and their improvements in the next decade.

The crude petroleum, of course, is not fit for illuminating purposes, but it contains components which are satisfactory. The various components are sorted out by fractional distillation and the oil for burning in lamps is selected according to its volatility, viscosity, stability, etc. It must not be so volatile as to have a dangerously low flashing-point, nor so stable as to hinder its burning well. In this fractional distillation a vast variety of products are now obtained. Gasolene is among the lighter products, with a density of about 0.65; kerosene has a density of about 0.80; the lubricating-oils from 0.85 to 0.95; and there are many solids such as vaseline and paraffin which are widely used for many purposes. This process of refining oils is now the source of paraffin for making candles, in which it is usually mixed with substances like stearin in order to raise its melting-point.

Crude petroleum possesses a very repugnant odor; it varies in color from yellow to black; and its specific gravity ranges from about 0.80 to 1.00, but commonly is between 0.80 and 0.90. Its chemical constitution is chiefly of carbon and hydrogen, in the approximate ratio of about six to one respectively. It is a mixture of paraffin hydrocarbons having the general formula of C_nH_{2n+2} and the individual members of this series vary from CH_4 (methane) to $C_{15}H_{32}$ (pentadecane), although the solid hydrocarbons are still more complex. Petroleum is found in many countries and the United States is particularly blessed with great stores of it.

The ordinary lamp consisting of a wick which draws up the mineral oil and feeds it to a flame is efficient and fairly free from danger. It requires care and may cause disaster if it is upset, but it has been blamed unjustly in many accidents. A disadvantage of the kerosene lamp over electric lighting, for example, is the relatively greater possibility of accidents through the carelessness of the user. This point is brought out in statistics of fire-insurance companies, which show that the fires caused by kerosene lamps are much more numerous than those from other methods of lighting. If in a modern lamp of proper construction a close-fitting wick is used and the lamp is extinguished by turning down and blowing across the chimney, there is little danger in its use excepting accidental breakage or overturning.

In oil-lamps at the present time mineral oils are used which possess flashing-points above 75 癒. The highly volatile components of petroleum are dangerous because they form very explosive mixtures with air at ordinary temperatures. A mineral oil like kerosene, to be used with safety in lamps, should not be too volatile. It is preferable that an inflammable vapor should not be given off at temperatures under 120. The oil must be of such physical characteristics as to be drawn up to the burner by capillarity from the reservoir which is situated below. It is volatilized by the heat of the flame into a mixture of hydrogen and hydrocarbon gases and these are consumed under the heat of the process of consumption by the oxygen in the air. The resulting products of this combustion, if it is complete, are carbon dioxide and water-vapor. For each candle-power of light per hour about 0.24 cubic foot of carbon dioxide and 0.18 cubic foot of water-vapor are formed by a modern oil-lamp. That an open flame devours something from the air is easily demonstrated by enclosing it in an air-tight space. The flame gradually becomes feeble and smoky and finally goes out. It will be noted that a burning lamp will vitiate the atmosphere of a closed room by consuming the oxygen and returning in its place carbon dioxide. This is similar to the vitiation of the atmosphere by breathing persons and tests indicate that for each two candle-power emitted by a kerosene flame the vitiation is equal to that produced by one adult person. Inasmuch as oil-lamps are ordinarily of 10 to 20 candle-power, it is seen that one lamp will consume as much oxygen as several persons.

In order that oil-lamps may produce a brilliant light free from smoke, combustion must be complete. The correct quantity of oil must be fed to the burner and it must be properly vaporized by heat. If insufficient oil is fed, the intensity of the light is diminished and if too much is available at the burner, smoke and other products of incomplete combustion will be emitted. The wick is an important factor, for, through capillarity, it feeds oil forcefully to the burner against the action of gravity. This action of a wick is commonly looked upon with indifference but in reality it is caused by an interesting and really wonderful phenomenon. Wicks are usually made of high-grade cotton fiber loosely spun into coarse threads and these are woven into a loose plait. The wick must be dry before being inserted into the burner; and it is desirable that it be considerably longer than is necessary merely to reach the bottom of the reservoir. A flame burning in the open will smoke because insufficient oxygen is brought in contact with it. The injurious products of this incomplete combustion are carbon monoxide and oil vapors, which are a menace to health.

To supply the necessary amount of oxygen (air) to the flame, a forced draft is produced. Chimneys are simple means of accomplishing this, and this is their function whether on oil-lamps or factories. Other means of forced draft have been used, such as small fans or compressed air. In the railway locomotive the short smoke-stack is insufficient for supplying large quantities of air to the fire-box so the exhausted steam is allowed to escape into the stack. With each noisy puff of smoke a quantity of air is forcibly drawn into the fire-box through the burning fuel. In the modern oil-lamp the rush of air due to the "pull" of the chimney is broken and the air is diffused by the wire gauze or holes at the base of the burner. These metal parts, being hot, also serve to warm the oil before it reaches the burning end of the wick, thus serving to aid vaporization and combustion.

The consumption of oil per candle-power per hour varies considerably with the kind of lamp and with the character of the oil. The average consumption of oil-lamps burning a mineral oil of about 0.80 specific gravity and a rather high flashing-point is about 50 to 60 grams of oil per candle-power per hour for well-designed flame-lamps. Kerosene weighs about 6.6 pounds per gallon; therefore, about 800 candle-power hours per gallon are obtained from modern lamps employing wicks. Kerosene lamps are usually of 10 to 20 candle-power, although they are made up to 100 candle-power. These luminous intensities refer to the maximum horizontal candle-power. The best practice now deals with the total light output, which is expressed in lumens, and on this basis a consumption of one gallon of kerosene per hour would yield about 8000 lumens.

Oil-lamps have been devised in which the oil is burned as a spray ejected by air-pressure. These burn with a large flame; however, a serious feature is the escape of considerable oil which is not burned. These lamps are used in industrial lighting, especially outdoors, and possess the advantage of consuming low-grade oils. They produce about 700 to 800 candle-power hours per gallon of oil. Lamps of this type of the larger sizes burn with vertical flames two or three feet high. The oil is heated as it approaches the nozzle and is fairly well vaporized on emerging into the air. The names of Lucigen, Wells, Doty, and others are associated with this type of lamp or torch, which is a step in the direction of air-gas lighting.

During the latter part of the nineteenth century numerous developments were made which paralleled the progress in gas-lighting. Experiments were conducted which bordered closely upon the next epochal event in light-production--the appearance of the gas mantle. One of these was the use of platinum gauze by Kitson. He produced an apparatus similar to the oil-spray lamp, on a small and more delicate scale. The hot blue flame was not very luminous and he attempted to obtain light by heating a mantle of fine platinum gauze. Although these mantles emitted a brilliant light for a few hours, their light-emissivity was destroyed by carbonization. After the appearance of the Welsbach mantle, Kitson's lamp and others met with success by utilizing it. From this point, attention was centered upon the new wonder, which is discussed in a later chapter after certain scientific principles in light-production have been discussed.

The kerosene or mineral-oil lamp was a boon to lighting in the nineteenth century and even to-day it is a blessing in many homes, especially in villages, in the country, and in the remote districts of civilization. Its extensive use at the present time is shown by the fact that about eight million lamp-chimneys are now being manufactured yearly in this country. It is convenient and safe when carelessness is avoided, and is fairly free from odor. Its vitiation of the atmosphere may be counteracted by proper ventilation and there remains only the disadvantage of keeping it in order and of accidental breakage and overturning. The kerosene lantern is widely used to-day, but the danger due to accident is ever-present. The consequences of such accidents are often serious and are exemplified in the terrible conflagration in Chicago in 1871, when Mrs. O'Leary's cow kicked over a lantern and started a fire which burned the city. Modern developments in lighting are gradually encroaching upon the territory in which the oil-lamp has reigned supreme for many years. Acetylene plants were introduced to a considerable extent some time ago and to-day the self-contained home-lighting electric plant is being installed in large numbers in the country homes of the land.

VI

EARLY GAS-LIGHTING

Owing to the fact that the smoky, flickering oil-lamp persisted throughout the centuries and until the magic touch of Argand in the latter part of the eighteenth century transformed it into a commendable light-source, the reader is prepared to suppose that gas-lighting is of recent origin. Apparently William Murdock in England was the first to install pipes for the conveyance of gas for lighting purposes. In an article in the "Philosophical Transactions of the Royal Society of London" dated February 25, 1808, in which he gives an account of the first industrial gas-lighting, he states:

It is now nearly sixteen years, since, in a course of experiments I was making at Redruth in Cornwall, upon the quantities and qualities of the gases produced by distillation from different mineral and vegetable substances, I was induced by some observation I had previously made upon the burning of coal, to try the combustible property of the gases produced from it....

Inasmuch as he is credited with having lighted his home by means of piped gas, this experimental installation may be considered to have been made in 1792. In his first trial he burned the gas at the open ends of the pipes; but finding this wasteful, he closed the ends and in each bored three small holes from which the gas-flames diverged. It is said that he once used his wife's thimble in an emergency to close the end of the pipe; and, the thimble being much worn and consequently containing a number of small holes, tiny gas-jets emerged from the holes. This incident is said to have led to the use of small holes in his burners. He also lighted a street lamp and had bladders filled with gas "to carry at night, with which, and his little steam carriage running on the road, he used to astonish the people." Apparently unknown to Murdock, previous observations had been made as to the inflammability of gas from coal. Long before this Dr. Clayton described some observations on coal-gas, which he called "the spirit of coals." He filled bladders with this gas and kept them for some time. Upon his pricking one of them with a pin and applying a candle, the gas burned at the hole. Thus Clayton had a portable gas-light. He was led to experiment with distillation of coal from some experiences with gas from a natural coal bed, and he thus describes his initial laboratory experiment:

I got some coal, and distilled it in a retort in an open fire. At first there came over only phlegm, afterwards a black oil, and then likewise, a spirit arose which I could no ways condense; but it forced my lute and broke my glasses. Once when it had forced my lute, coming close thereto, in order to try to repair it, I observed that the spirit which issued out caught fire at the flame of the candle, and continued burning with violence as it issued out in a stream, which I blew out, and lighted again alternately several times.

He then turned his attention to saving some of the gas and hit upon the use of bladders. He was surprised at the amount of gas which was obtained from a small amount of coal; for, as he stated, "the spirit continued to rise for several hours, and filled the bladders almost as fast as a man could have blown them with his mouth; and yet the quantity of coals distilled was inconsiderable."

Although this account appeared in the Transactions of the Royal Society in 1739, there is strong evidence that Dr. Clayton had written it many years before, at least prior to 1691.

But before entering further into the early history of gas-lighting, it is interesting to inquire into the knowledge possessed in the seventeenth century pertaining to natural and artificial gas. Doubtless there are isolated instances throughout history of encounters with natural gas. Surely observant persons of bygone ages have noted a small flame emanating from the end of a stick whose other end was burning in a bonfire or in the fireplace. This is a gas-plant on a small scale; for the gas is formed at the burning end of the wooden stick and is conducted through its hollow center to the cold end, where it will burn if lighted. If a piece of paper be rolled into the form of a tube and inclined somewhat from a horizontal position, inflammable gas will emanate from the upper end if the lower end is burning. By applying a match near the upper end, we can ignite this jet of gas. However, it is certain that little was known of gas for illuminating purposes before the eighteenth century.

The literature of an ancient nation is often referred to as revealing the civilization of the period. Surely the scientific literature which deals with concrete facts is an exact indicator of the technical knowledge of a period! That little was known of natural gas and doubtless of artificial gas in the seventeenth century is shown by a brief report entitled "A Well and Earth in Lancashire taking Fire at a Candle," by Tho. Shirley in the Transactions of the Royal Society in 1667. Much of the quaint charm of the original is lost by inability to present the text in its original form, but it is reproduced as closely as practicable. The report was as follows:

About the latter End of Feb. 1659, returning from a Journey to my House in Wigan, I was entertained with the Relation of an odd Spring situated in one

Mr. Hawkley's Ground (if I mistake not) about a Mile from the Town, in that Road which leads to Warrington and Chester: The People of this Town did confidently affirm, That the Water of this Spring did burn like Oil.

When we came to the said Spring (being 5 or 6 in Company together) and applied a lighted Candle to the Surface of the Water; there was 'tis true, a large Flame suddenly produced, which burnt the Foot of a Tree, growing on the Top of a neighbouring Bank, the Water of which Spring filled a Ditch that was there, and covered the Burning-place; I applied the lighted Candle to divers Parts of the Water contained in the said Ditch, and found, as I expected, that upon the Touch of the Candle and the Water the Flame was extinct.

Again, having taken up a Dish full of water at the flaming Place, and held the lighted Candle to it, it went out. Yet I observed that the Water, at the Burning-place, did boil, and heave, like Water in a Pot upon the Fire, tho' by putting my Hand into it, I could not perceive it so much as warm.

This Boiling I conceived to proceed from the Eruption of some bituminous or sulphureous Fumes; considering this Place was not above 30 or 40 Yards distant from the Mouth of a Coal-Pit there: And indeed Wigan, Ashton, and the whole Country, for many Miles compass, is underlaid with Coal. Then, applying my Hand to the Surface of the Burning-place of the Water, I found a strong Breath, as it were a Wind, to bear against my Hand.

When the Water was drained away, I applied the Candle to the Surface of the dry Earth, at the same Point where the Water burned before; the Fumes took fire, and burned very bright and vigorous. The Cone of the Flame ascended a Foot and a half from the Superficies of the Earth; and the Basis of it was of the Compass of a Man's Hat about the Brims. I then caused a Bucket full of Water to be pour'd on the Fire, by which it was presently quenched. I did not perceive the Flame to be discoloured like that of sulphurous Bodies, nor to have any manifest Scent with it. The Fumes, when they broke out of the Earth, and press'd against my Hand, were not, to my best Remembrance, at all hot.

Turning again to Dr. Clayton's experiments, we see that he pointed out striking and valuable properties of coal-gas but apparently gave no attention to its useful purposes. Furthermore, his account appears to have attracted no particular notice at the time of its publication in 1739. Dr. Richard Watson in 1767 described the results of experiments which he had been making with the products arising from the distillation of coal. In his process he permitted the gas to ascend through curved tubes, and he particularly noted "its great inflammability as well as elasticity." He also observed that "it retained the former property after it had passed through a great quantity of water." His published account dealt with a variety of facts and computations pertaining to the quantities of coke, tar, etc., produced from different kinds of coal and was a scientific work of value, but apparently the usefulness of the property of inflammability of coal-gas did not occur to him.

It is usually the habit of the scientific explorer of nature to return from excursions into her unfrequented recesses with new knowledge, to place it upon exhibition, and to return for more. The inventor passes by and sees applications for some of these scientific trophies which are productive of momentous consequences to mankind. Sir Humphrey Davy described his primitive arc-lamp three quarters of a century before Brush developed an arc-lamp for practical purposes. Maxwell and Hertz respectively predicted and produced electromagnetic waves long before Marconi applied this knowledge and developed "wireless" telegraphy. In a similar manner scientific accounts of the production and properties of coal-gas antedated by many years the initial applications made by Murdock to illuminating purposes.

Up to the beginning of the nineteenth century the civilized world had only a faint glimpse of the illuminating property of gas, but practicable gas-lighting was destined soon to be an epochal event in the progress of lighting. The dawn of modern science was coincident with the dawn of a luminous era.

At Soho foundry in 1798 Murdock constructed an apparatus which enabled him to exhibit his lighting-plan on a larger scale and to experiment on purifying and burning the gas so as to eliminate odor and smoke. Soho was an unique institution described as a place

to which men of genius were invited and resorted from every civilized country, to exercise and to display their talents. The perfection of the manufacturing arts was the great and constant aim of its liberal and enlightened proprietors, Messrs. Boulton and Watt; and whoever resided there was surrounded by a circle of scientific, ingenious, and skilful men, at all times ready to carry into effect the inventions of each other.

The Treaty of Amiens, which gave to England the peace she was sorely in need of, afforded Murdock an opportunity in 1802 favorable for making a public display of gas-lighting. The illumination of the Soho works on this occasion is described as "one of extraordinary splendour." The fronts of the extensive range of buildings were ornamented with a large number of devices which displayed the variety of forms of gas-lights. At that time this was a luminous spectacle of great novelty and the populace came from far and wide "to gaze at, and to admire, this wonderful display of the combined effects of science and art."

Naturally, Murdock had many difficulties to overcome in these early days, but he possessed skill and perseverance. His first retorts for distilling coal were similar to the common glass retort of the chemist. Next he tried cast-iron cylinders placed perpendicularly in a common furnace, and in each were put about fifteen pounds of coal. In 1804 he constructed them with doors at each end, for feeding coal and extracting coke respectively, but these were found inconvenient. In his first lighting installation in the factory of Phillips and Lee in 1805 he used a large retort of the form of a bucket with a cover on it. Inside he installed a loose cage of grating to hold the coal. When carbonization was complete the coke could be removed as a whole by extracting this cage. This retort had a capacity of fifteen hundred pounds of coal. He labored with mechanical details, varied the size and shape of the retorts, and experimented with different temperatures, with the result that he laid a solid foundation for coal-gas lighting. For his achievements he is entitled to an honorable place among the torch-bearers of civilization.

The epochal feature of the development of gas-lighting is that here was a possibility for the first time of providing lighting as a public utility. In the early years of the nineteenth century the foundation was laid for the great public-utility organizations of the present time. Furthermore, gas-lighting was an improvement over candles and oil-lamps from the standpoints of convenience, safety, and cost. The latter points are emphasized by Murdock in his paper presented before the Royal Society in 1808, in which he describes the first industrial installation of gas-lighting. He used two types of burners, the Argand and the cockspur. The former resembled the Argand lamp in some respects and the latter was a three-flame burner suggesting a fleur-delis. In this installation there were 271 Argand burners and 636 cockspurs. Each of the former "gave a light equal to that of four candles; and each of the latter, a light equal to two and a quarter of the same candles; making therefore the total of the gas light a little more than 2500 candles." The candle to which he refers was a mold candle "of six in the pound" and its light was considered a standard of luminous intensity when it was consuming tallow at the rate of 0.4 oz. (175 grains) per hour. Thus the candle became very early a standard light-source and has persisted as such (with certain variations in the specifications) until the present time. However, during recent years other standard light-sources have been devised.

According to Murdock, the yearly cost of gas-lighting in this initial case was 600 pounds sterling after allowing generously for interest on capital invested and depreciation of the apparatus. The cost of furnishing the same amount of light by means of candles he computed to be 2000 pounds sterling. This comparison was on the basis of an average of two hours of artificial lighting per day. On the basis of three hours of artificial lighting per day, the relative cost of gas-and candle-lighting was about one to five. Murdock was characteristically modest in discussing his achievements and his following statement should be read with the conditions of the year 1808 in mind:

The peculiar softness and clearness of this light with its almost unvarying intensity, have brought it into great favour with the work people. And its being free from the inconvenience and danger, resulting from sparks and frequent snuffing of candles, is a circumstance of material importance, as tending to diminish the hazard of fire, to which cotton mills are known to be exposed.

Although this installation in the mill of Phillips and Lee is the first one described by Murdock, in reality it is not the first industrial gas-lighting installation. During the development of gas apparatus at the Soho works and after his luminous display in 1802, he gradually extended gas-lighting to all the principal shops. However, this in a sense was experimental work. Others were applying their knowledge and ingenuity to the problem of making gas-lighting practicable, but Murdock has been aptly termed "the father of gas-lighting." Among the pioneers was Le Bon in France, Becher in Munich, and Winzler or Winsor, a German who was attracted to the possibilities of gas-lighting by an exhibition which Le Bon gave in Paris in 1802. Winsor learned that Le Bon had been granted a patent in Paris in 1799 for making an illuminating gas from wood and tried to obtain the rights for Germany. Being unsuccessful in this, he set about to learn the secrets of Le Bon's process, which he did, perhaps largely owing to an accumulation of information directly from the inventor during the negotiations. Winsor then turned to England as a fertile field for the exploitation of gas-lighting and after conducting experiments in London for some time he made plans to organize the National Heat and Light Co.

Winsor was primarily a promoter, with little or no technical knowledge; for in his claims and advertisements he disregarded facts with a facility possessed only by the ignorant. He boasted of his inventions and discoveries in the most hyperbolical language, which was bound to provoke a controversy. Nevertheless, he was clever and in 1803 he publicly exhibited his plan of lighting by means of coal-gas at the Lyceum Theatre in London. He gave lectures accompanied by interesting and instructive experiments and in this manner attracted the public to his exhibition. All this time he was promoting his company, but his promoting instinct caused his representations to be extravagant and deceptive, which exposed him to the ridicule and suspicion of learned men. His attempt to obtain certain exclusive rights by Act of Parliament failed because of opposition of scientific men toward his claims and of the stand which Murdock justly made in self-protection. These years of controversy yield entertaining literature for those who choose to read it, but unfortunately space does not permit dwelling upon it. The investigations by committees of Parliament also afford amusing side-lights. Throughout all this Murdock appeared modest and conservative and had the support of reputable scientific men, but Winsor maintained extravagant claims.

During one of these investigations Sir Humphrey Davy was examined by a committee from the House of Commons in 1809. He refuted Winsor's claims for a superior coke as a by-product and stated that the production of gas by the distillation of coal had been well known for thirty or forty years and the production of tar as long. He stated that it was the opinion of the Council of the Royal Society that Murdock was the first person to apply coal-gas to lighting in actual practice. As secretary of the Society, Sir Humphrey Davy stated that at the last session it had bestowed the Count Rumford medal upon Murdock for "his economical application of the gas light."

Winsor proceeded to float his company without awaiting the Act of Parliament and in 1807 lighted a street in Pall Mall. Through the opposition which he aroused, and owing to the just claims of priority on the part of Murdock, the bill to incorporate the National Heat and Light Co. with a capital of 200,000 pounds sterling was thrown out. However, he succeeded in 1812 in receiving a charter very much modified in form, for the Chartered Gas Light and Coke Co. which was the forerunner of the present London Gas Light and Coke Co.

The conditions imposed upon this company as presented in an early treatise on gas-lighting (by Accum in 1818) were as follows:

The power and authorities granted to this corporate body are very restricted and moderate. The individuals composing it have no exclusive privilege; their charter does not prevent other persons from entering into competition with them. Their operations are confined to the metropolis, where they are bound to furnish not only a stronger and better light to such streets and parishes as chuse to be lighted with gas, but also at a cheaper price than shall be paid for lighting the said streets with oil in the usual manner. The corporation is not permitted to traffic in machinery for manufacturing or conveying the gas into private houses, their capital or joint stock is limited to ?00,000, and his Majesty has the power of declaring the gas-light charter void if the company fail to fulfil the terms of it.

The progress of this early company was slow at first, but with the appointment of Samuel Clegg as engineer in 1813 an era of technical developments began. New stations were built and many improvements were introduced. By improving the methods of purifying the gas a great advance was made. The utility of gas-lighting grew apace as the prejudices disappeared, but for a long time the stock of the company sold at a price far below par. About this time the first gas explosion took place and the members of the Royal Society set a precedent which has lived and thrived: they appointed a committee to make an inquiry. But apparently the inquiry was of some value, for it led "to some useful alterations and new modifications in its apparatus and machinery."

Many improvements were being introduced during these years and one of them in 1816 increased the gaseous product from coal by distilling the tar which was obtained during the first distillation. In 1816 Clegg obtained a patent for a horizontal rotating retort; for an apparatus for purifying coal-gas with "cream of lime"; and for a rotative gas-meter.

Before progressing too far, we must mention the early work of William Henry. In 1804 he described publicly a method of producing coal-gas. Besides making experiments on production and utilization of coal-gas for lighting, he devoted his knowledge of chemistry to the analysis of the gas. He also made analytical studies of the relative value of wood, peat, oil, wax, and different kinds of coal for the distillation of gas. His chemical analyses showed to a considerable extent the properties of carbureted hydrogen upon which illuminating value depended. The results of his work were published in various English journals between 1805 and 1825 and they contributed much to the advancement of gas-lighting.

Although Clegg's original gas-meter was complicated and cumbersome, it proved to be a useful device. In fact, it appears to have been the most original and beneficial invention occasioned by early gas-lighting. Later Samuel Crosley greatly improved it, with the result that it was introduced to a considerable extent; but by no means was it universally adopted. Another improvement made by Clegg at this time was a device which maintained the pressure of gas approximately constant regardless of the pressure in the gasometer or tank. Clegg retired from the service of the gas company in 1817 after a record of accomplishments which glorifies his name in the annals of gas-lighting. Murdock is undoubtedly entitled to the distinction of having been the first person who applied gas-lighting to large private establishments, but Clegg overcame many difficulties and was the first to illuminate a whole town by this means.

In London in 1817 over 300,000 cubic feet of coal-gas was being manufactured daily, an amount sufficient to operate 76,500 Argand burners yielding 6 candle-power each. Gas-lighting was now exciting great interest and was firmly established. Westminster Bridge was lighted by gas in 1813, and the streets of Westminster during the following year. Gas-lighting became popular in London by 1816 and in the course of the next few years it was adopted by the chief cities and towns in the United Kingdom and on the Continent. It found its way into the houses rather slowly at first, owing to apprehension of the attendant dangers, to the lack of purification of the gas, and to the indifferent service. It was not until the latter half of the nineteenth century that it was generally used in residences.

The gas-burner first employed by Murdock received the name "cockspur" from the shape of the flame. This had an illuminating value equivalent to about one candle for each cubic foot of gas burned per hour. The next step was to flatten the welded end of the gas-pipe and to bore a series of holes in a line. From the shape of the flames this form of burner received the name "cockscomb." It was somewhat more efficient than the cockspur burner. The next obvious step was to slit the end of the pipe by means of a fine saw. From this slit the gas was burned as a sheet of flame called the "bats-wing." In 1820 Nielson made a burner which allowed two small jets to collide and thus form a flat flame. The efficiency of this "fish-tail" burner was somewhat higher than that of the earlier ones. Its flame was steadier because it was less influenced by drafts of air. In 1853 Frankland showed an Argand burner consisting of a metal ring containing a series of holes from which jets of gas issued. The glass chimney surrounded these, another chimney, extending somewhat lower, surrounded the whole, and a glass plate closed the bottom. The air to be fed to the gas-jets came downward between the two chimneys and was heated before it reached the burner. This increased the efficiency by reducing the amount of cooling at the burner by the air required for combustion. This improvement was in reality the forerunner of the regenerative lamps which were developed later.

In 1854 Bowditch brought out a regenerative lamp and, owing to the excessive publicity which this lamp obtained, he is generally credited with the inception of the regenerative burner. This principle was adopted in several lamps which came into use later. They were all based upon the principle of heating both the gas and the air required for combustion prior to their reaching the burner. The burner is something like an inverted Argand arranged to produce a circular flame projecting downward with a central cusp. The air- and gas-passages are directly above the flame and are heated by it. In 1879 Friedrich Siemens brought out a lamp of this type which was adapted from a device originally designed for heating purposes, owing to the superior light which was produced. This was the best gas-lamp up to that time. Later, Wenham, Cromartie, and others patented lamps operating on this same principle.

Murdock early modified the Argand burner to meet the requirements of burning gas and by using the chimney obtained better combustion and a steadier flame than from the open burners. He and others recognized that the temperature of the flame had a considerable effect upon the amount of light emitted and non-conducting material such as steatite was substituted for the metal, which cooled the flame by conducting heat from it. These were the early steps which led finally to the regenerative burner.

The increasing efficiency of the various gas-burners is indicated by the following, which are approximately the candle-power based upon equal rates of consumption, namely, one cubic foot of gas per hour:

Candle-power per cubic foot of gas per hour

Fish-tail flames, depending upon size 0.6 to 2.5 Argand, depending upon improvements 2.9 to 3.5 Regenerative 7 to 10

It is seen that the possibilities of gas lighting were recognized in several countries, all of which contributed to its development. Some of the earlier accounts have been drawn chiefly from England, but these are intended merely to serve as examples of the difficulties encountered. Doubtless, similar controversies arose in other countries in which pioneers were also nursing gas-lighting to maturity. However, it is certain that much of the early progress of lighting of this character was fathered in England. Gas-lighting was destined to become a thriving industry, and is of such importance in lighting that another chapter is given its modern developments.

VII

THE SCIENCE OF LIGHT-PRODUCTION

In previous chapters much of the historical development of artificial lighting has been presented and several subjects have been traced to the modern period which marks the beginning of an intensive attack by scientists upon the problems pertaining to the production of efficient and adequate light-sources. Many historical events remain to be touched upon in later chapters, but it is necessary at this point for the reader to become acquainted with certain general physical principles in order that he may read with greater interest some of the chapters which follow. It is seen that from a standpoint of artificial lighting, the "dark age" extended well into the nineteenth century. Oil-lamps and gas-lighting began to be seriously developed at the beginning of the last century, but the pioneers gave attention chiefly to mechanical details and somewhat to the chemistry of the fuels. It was not until the science of physics was applied to light-sources that rapid progress was made.

All the light-sources used throughout the ages, and nearly all modern ones, radiate light by virtue of the incandescence of solids or of solid particles and it is an interesting fact that carbon is generally the solid which emits light. This is due to various physical characteristics of carbon, the chief one being its extremely high melting-point. However, most practicable light-sources of the past and present may be divided into two general classes: (1) Those in which solids or solid particles are heated by their own combustion, and (2) those in which the solids are heated by some other means. Some light-sources include both principles and some perhaps cannot be included under either principle without qualification. The luminous flames of burning material such as those of wood-splinters, candles, oil-lamps, and gas-jets, and the glowing embers of burning material appear in the first class; and incandescent gas-mantles, electric filaments, and arc-lamps to some extent are representative of the second class. Certain "flaming" arcs involve both principles, but the light of the firefly, phosphorescence, and incandescent gas in "vacuum" tubes cannot be included in this simplified classification. The status of these will become clear later.

It has been seen that flames have been prominent sources of artificial light; and although of low luminous efficiency, they still have much to commend them from the standpoints of portability, convenience, and subdivision. The materials which have been burned for light, whether solid or liquid, are rich in carbon, and the solid particles of carbon by virtue of their incandescence are responsible for the brightness of a flame. A jet of pure hydrogen gas will burn very hot but with so low a brightness as to be barely visible. If solid particles are injected into the flame, much more light usually will be emitted. A gas-burner of the Bunsen type, in which complete combustion is obtained by mixing air in proper proportions with the gas, gives a hot flame which is of a pale blue color. Upon the closing of the orifice through which air is admitted, the flame becomes bright and smoky. The flame is now less hot, as indicated by the presence of smoke or carbon particles, and combustion is not complete. However, it is brighter because the solid particles of carbon in passing upward through the flame become heated to temperatures at which they glow and each becomes a miniature source of light.

A close observer will notice that the flame from a match, a candle, or a gas-jet, is not uniformly bright. The reader may verify this by lighting a match and observing the flame. There is always a bluish or darker portion near the bottom. In this less luminous part the air is combining with the hydrogen of the hydrocarbon which is being vaporized and disintegrated. Even the flame of a candle or of a burning splinter is a miniature gas-plant, for the solid or liquid hydrocarbons are vaporized before being burned. Owing to the incoming colder air at this point, the flame is not hot enough for complete combustion. The unburned carbon particles rise in its draft and become heated to incandescence, thus accounting for the brighter portion. In cases of complete combustion they are eventually oxidized into carbon dioxide before they are able to escape. If a piece of metal be held in the flame, it immediately becomes covered with soot or carbon, because it has reduced the temperature below the point at which the chemical reaction--the uniting of carbon with oxygen--will continue. An ordinary flat gas-flame of the "bats-wing" type may vary in temperature in its central portion from 300. at the bottom to about 3000. at the top. The central portion lies between two hotter layers in which the vertical variation is not so great. The brightness of the upper portion is due to incandescent carbon formed in the lower part.

When scientists learned by exploring flames that brightness was due to the radiation of light by incandescent solid matter, the way was open for many experiments. In the early days of gas-lighting investigations were made to determine the relation of illuminating value to the chemical constitution of the gas. The results combined with a knowledge of the necessity for solid carbon in the flame led to improvements in the gas for lighting purposes. Gas rich in hydrocarbons which in turn are rich in carbon is high in illuminating value. Heating-effect depends upon heat-units, so the rating of gas in calories or other heat-units per cubic foot is wholly satisfactory only for gas used for heating. The chemical constitution is a better indicator of illuminating value.

As scientific knowledge increased, efforts were made to get solid matter into the flames of light-sources. Instead of confining efforts to the carbon content of the gas, solid materials were actually placed in the flame, and in this manner various incandescent burners were developed. A piece of lime placed in a hydrogen flame or that of a Bunsen burner is seen to become hot and to glow brilliantly. By producing a hotter flame by means of the blowpipe, in which hydrogen and oxygen are consumed, the piece of lime was raised to a higher temperature and a more intense light was obtained. In Paris there was a serious attempt at street-lighting by the use of buttons of zirconia heated in an oxygen-coal-gas flame, but it proved unsuccessful owing to the rapid deterioration of the buttons. This was the line of experimentation which led to the development of the lime-light. The incandescent burner was widely employed, and until the use of electricity became common the lime-light was the mainstay for the stage and for the projection of lantern slides. It is in use even to-day for some purposes. The origin of the phrase "in the lime-light" is obvious. The luminous intensity of the oxyhydrogen lime-light as used in practice was generally from 200 to 400 candle-power. The light decreases rapidly as the burner is used, if a new surface of lime is not presented to the flame from time to time. At the high temperatures the lime is somewhat volatile and the surface seems to change in radiating power. Zirconium oxide has been found to serve better than lime.

Improvements were made in gas-burners in order to obtain hotter flames into which solid matter could be introduced to obtain bright light. Many materials were used, but obviously they were limited to those of a fairly high melting-point. Lime, magnesia, zirconia, and similar oxides were used successfully. If the reader would care to try an experiment in verification of this simple principle, let him take a piece of magnesium ribbon such as is used in lighting for photography and ignite it in a Bunsen flame. If it is held carefully while burning, a ribbon of ash (magnesia) will be obtained intact. Placing this in the faintly luminous flame, he will be surprised at the brilliance of its incandescence when it has become heated. The simple experiment indicates the possibilities of light-production in this direction. Naturally, metals of high melting-point such as platinum were tried and a network of platinum wire, in reality a platinum mantle, came into practical use in about 1880. The town of Nantes was lighted by gas-burners using these platinum-gauze mantles, but the mantles were unsuccessful owing to their rapid deterioration. This line of experimentation finally bore fruit of immense value for from it the gas-mantle evolved.

A group of so-called "rare-earths," among which are zirconia, thoria, ceria, erbia, and yttria (these are oxides of zirconium, etc.) possess a number of interesting chemical properties some of which have been utilized to advantage in the development of modern artificial light. They are white or yellowish-white oxides of a highly refractory character found in certain rare minerals. Most of them are very brilliant when heated to a high temperature. This latter feature is easily explained if the nature of light and the radiating properties of substances are considered. Suppose pieces of different substances, for example, glass and lime, are heated in a Bunsen flame to the same temperature which is sufficiently great to cause both of them to glow. Notwithstanding the identical conditions of heating, the glass will be only faintly luminous, while the piece of lime will glow brilliantly. The former is a poor radiator; furthermore, the lime radiates a relatively greater percentage of its total energy in the form of luminous energy.

The latter point will become clearer if the reader will refresh his memory regarding the nature of light. Any luminous source such as the sun, a candle flame, or an incandescent lamp is sending forth electromagnetic waves not unlike those used in wireless telegraphy excepting that they are of much shorter wave-length. The eye is capable of recording some of these waves as light just as a receiving station is tuned to record a range of wave-lengths of electromagnetic energy. The electromagnetic waves sent forth by a light-source like the sun are not all visible, that is, all of them do not arouse a sensation of light. Those that do comprise the visible spectrum and the different wave-lengths of visible radiant energy manifest themselves by arousing the sensations of the various spectral colors. The radiant energy of shortest wave-length perceptible by the visual apparatus excites the sensation of violet and the longest ones the sensation of deep red. Between these two extremes of the visible spectrum, the chief spectral colors are blue, green, yellow, orange, and red in the order of increasing wave-lengths. Electromagnetic energy radiated by a light-source in waves of too great wave-length to be perceived by the eye as light is termed as a class "infra-red radiant energy." Those too short to be perceived as light are termed as a class "ultraviolet radiant energy." A solid body at a high temperature emits electro-magnetic energy of all wave-lengths, from the shortest ultra-violet to the longest infra-red.

Another complication arises in the variation in visibility or luminosity of energy of wave-lengths within the range of the visible spectrum. Obviously, no amount of energy incapable of exciting the sensation of light will be visible. The energy of those wave-lengths near the ends of the visible spectrum will be inefficient in producing light. That energy which excites the sensation of yellow-green produces the greatest luminosity per unit of energy and is the most efficient light. The visibility or luminous efficiency of radiant energy may be ranged approximately in this manner according to the colors aroused: yellow-green, yellow, green, orange, blue-green, red, blue, deep red, violet.

Newton, an English scientist, first described the discovery of the visible spectrum and this is of such fundamental importance in the science of light that the first paragraph of his original paper in the "Transactions of the Royal Society of London" is quoted as follows:

In the Year 1666 (at which time I applied my self to the Grinding of Optick Glasses of other Figures than Spherical) I procured me a Triangular Glass-Prism, to try therewith the celebrated Phaenomena of Colours. And in order thereto, having darkened my Chamber, and made a small Hole in my Window-Shuts, to let in a convenient Quantity of the Sun's Light, I placed my Prism at its Entrance, that it might be thereby refracted to the opposite Wall. It was at first a very pleasing Divertisement, to view the vivid and intense Colours produced thereby; but after a while applying my self to consider them more circumspectly, I became surprised to see them in an oblong Form; which, according to the receiv'd Law of Refractions, I expected should have been circular. They were terminated at the Sides with streight Lines, but at the Ends the Decay of Light was so gradual, that it was difficult to determine justly what was the Figure, yet they seemed Semicircular.

Even Newton could not have had the faintest idea of the great developments which were to be based upon the spectrum.

Now to return to the peculiar property of rare-earth oxides--namely, their unusual brilliance when heated in a flame--it is easy to understand the reason for this. For example, when a number of substances are heated to the same temperature they may radiate the same amount of energy and still differ considerably in brightness. Many substances are "selective" in their absorbing and radiating properties. One may radiate more luminous energy and less infra-red energy, and for another the reverse may be true. The former would appear brighter than the latter. The scientific worker in light-production has been searching for such "selective" radiators whose other properties are satisfactory. The rare-earths possess the property of selectivity and are fortunately highly refractory. Welsbach used these in his mantle, whose efficiency is due partly to this selective property. Recent work indicates that much higher efficiencies of light-production are still attainable by the principles involved in the gas-mantle.

Turning again to flames, another interesting physical phenomenon is seen on placing solutions of different chemical salts in the flame. For example, if a piece of asbestos is soaked in sodium chloride (common salt) and is placed in a Bunsen flame, the pale-blue flame suddenly becomes luminous and of a yellow color. If this is repeated with other salts, a characteristic color will be noted in each case. The yellow flame is characteristic of sodium and if it is examined by means of a spectroscope, a brilliant yellow line (in fact, a double line) will be seen. This forms the basis of spectrum analysis as applied in chemistry.

Every element has its characteristic spectrum consisting usually of lines, but the complexity varies with the elements. The spectra of elements also exhibit lines in the ultra-violet region which may be studied with a photographic plate, with a photo-electric cell, and by other means. Their spectral lines or bands also extend into the infra-red region and here they are studied by means of the bolometer or other apparatus for detecting radiant energy by the heat which it produces upon being absorbed. Spectrum analysis is far more sensitive than the finest weighing balance, for if a grain of salt be dissolved in a barrel of water and an asbestos strip be soaked in the water and held in a Bunsen flame, the yellow color characteristic of sodium will be detectable. A wonderful example of the possibilities of this method is the discovery of helium in the sun before it was found on earth! Its spectral lines were detected in the sun's spectrum and could not be accounted for by any known element. However, it should be stated that the spectrum of an element differs generally with the manner obtained. The electric spark, the arc, the electric discharge in a vacuum tube, and the flame are the means usually employed.

The spectrum has been dwelt upon at some length because it is of great importance in light-production and probably will figure strongly in future developments. Although in lighting little use has been made of the injection of chemical salts into ordinary flames, it appears certain that such developments would have risen if electric illuminants had not entered the field. However, the principle has been applied with great success in arc-lamps. In the first arc-lamps plain carbon electrodes were used, but in some of the latest carbon-arcs, electrodes of carbon impregnated with various salts are employed. For example, calcium fluoride gives a brilliant yellow light when used in the carbons of the "flame" arc. These are described in detail later.

Following this principle of light-production the vacuum tubes were developed. Crookes studied the light from various gases by enclosing them in a tube which was pumped out until a low vacuum was produced. On connecting a high voltage to electrodes in each end, an electrical discharge passed through the residual gas making it luminous. The different gases show their characteristic spectra and their desirability as light-producers is at once evident.

However, the most general principle of light-production at the present time is the radiation of bodies by virtue of their temperature. If a piece of wire be heated by electricity, it will become very hot before it becomes luminous. At this temperature it is emitting only invisible infra-red energy and has an efficiency of zero as a producer of light. As it becomes hotter it begins to appear red, but as its temperature is raised it appears orange, until if it could be heated to the temperature of the sun, about 10,000., it would appear white. All this time its luminous efficiency is increasing, because it is radiating not only an increasing percentage of visible radiant energy but an increasing amount of the most effective luminous energy. But even when it appears white, a large amount of the energy which it radiates is invisible infra-red and ultra-violet, which are ineffective in producing light, so at best the substance at this high temperature is inefficient as a light-producer.

In this branch of the science of light-production substances are sought not only for their high melting-point, but for their ability to radiate selectively as much visible energy as possible and of the most luminous character. However, at best the present method of utilizing the temperature radiation of hot bodies has limitations.

The luminous efficiencies of light-sources to-day are still very low, but great advances have been made in the past half-century. There must be some radical departures if the efficiency of light-production is to reach a much higher figure. A good deal has been said of the firefly and of phosphorescence. These light-sources appear to emit only visible energy and, therefore, are efficient as radiators of luminous radiant energy. But much remains to be unearthed concerning them before they will be generally applicable to lighting. If ultra-violet radiation is allowed to impinge upon a phosphorescent material, it will glow with a considerable brightness but will be cool to the touch. A substance of the same brightness by virtue of its temperature would be hot; hence phosphorescence is said to be "cold" light.

An acquaintance with certain terms is necessary if the reader is to understand certain parts of the text. The early candle gradually became a standard, and uniform candles are still satisfactory standards where high accuracy is not required. Their luminous intensity and illuminating value became units just as the foot was arbitrarily adopted as a unit of length. The intensity of other light-sources was represented in terms of the number of candles or fraction of a candle which gave the same amount of light. But the luminous intensity of the candle was taken only in the horizontal direction. In the same manner the luminous intensities of light-sources until a short time ago were expressed in candles as measured in a certain direction. Incandescent lamps were rated in terms of mean horizontal candles, which would be satisfactory if the luminous intensity were the same in all directions, but it is not. Therefore, the candle-power in one direction does not give a measure of the total light-output.

If a source of light has a luminous intensity of one candle in all directions, the illumination at a distance of one foot in any direction is said to be a foot-candle. This is the unit of illumination intensity. A lumen is the quantity of light which falls on one square foot if the intensity of illumination is one foot-candle. It is seen that the area of a sphere with a radius of one foot is 4 pi or 12.57 square feet; therefore, a light-source having a luminous intensity of one candle in all directions emits 12.57 lumens. This is the satisfactory unit, for it measures total quantity of light, and luminous efficiencies may be expressed in terms of lumens per watt, lumens per cubic foot of gas per hour, etc.

Of course, the efficiencies of light-sources are usually of interest to the consumer if they are expressed in terms of cost. But from a practical point of view there are many elements which combine to make another important factor, namely, satisfactoriness. Therefore, the efficiency of artificial lighting from the standpoint of the consumer should be the ratio of satisfactoriness to cost. However, the scientist is interested chiefly in the efficiency of the light-source which may be expressed in lumens per watt, or the amount of light obtained from a given rate of consumption or of emission of energy. This method of rating light-sources penalizes those radiating considerable energy which does not produce the sensation of light or which at best is of wave-lengths that are inefficient in this respect. That radiant energy which is wholly of a wave-length of maximum visibility, or, in other words, excites the sensation of yellow-green, is the most efficient in producing luminous sensation. Of course, no illuminants are available which approach this theoretical ideal and it is not likely that this would be a practical ideal. Under monochromatic yellow-green light the magical drapery of color would disappear and the surroundings would be a monochrome of shades of this hue. Having no colors with which to contrast this color, the world would be colorless. This should be obvious when it is considered that an object which is red under an illuminant containing all colors such as sunlight would be black or dark gray under monochromatic yellow-green light. The red under present conditions is kept alive by contrast with other colors, because the latter live by virtue of the fact that most of our present illuminants contain their hues. It is assumed that the reader knows that a red object, for example, appears red because it reflects (or transmits) red rays and absorbs the other rays in the illuminant. In other words, color is due to selective absorption reflection, or transmission.

Perhaps the ideal illuminant, which is most generally satisfactory for general activities, is a white light corresponding to noon sunlight. If this is chosen as the scientific ideal, the illuminants of the present time are much more "efficient" than if the most efficient light is the ideal.

The luminous efficiency of the radiant energy most efficient in producing the sensation of light (yellow-green) is about 625 lumens per watt. That is, if energy of this wave-length alone were radiated by a hypothetical light-source, each watt would produce 625 lumens. The luminous efficiency of the most efficient white light is about 265 lumens per watt; in other words, if a hypothetical light-source radiated energy of only the visible wave-lengths and in proportions to produce the sensation of white, each watt would produce 265 lumens. If such a white light were obtained by pure temperature radiation--that is, by a normal radiator at a temperature of 10,000., which is impracticable at present--the luminous efficiency would be about 100 lumens per watt. The normal radiator which emits energy by virtue of its temperature without selectively radiating more or less energy in any part of the spectrum than indicated by the theoretical radiation laws is called a "black-body" or normal radiator. Modern illuminants have luminous efficiencies ranging from 5 to 30 lumens per watt, so it is seen that much is to be done before the limiting efficiencies are reached.

The amount of light obtained from various gas-burners for each cubic foot of gas consumed per hour varies for open gas-flames from 5 to 30 lumens; for Argand burners from 35 to 40 lumens; for regenerative lamps from 50 to 75 lumens; and for gas-mantles from 200 to 250 lumens.

In the development of light-sources, of course, any harmful effects of gases formed by burning or chemical action must be avoided. Some of the fumes from arcs are harmful, but no commercial arc appears to be dangerous when used as it is intended to be used. Gas-burners rob the atmosphere of oxygen and vitiate it with gases, which, however, are harmless if combustion is complete. That adequate ventilation is necessary where oxygen is being consumed is evident from the data presented by authorities on hygiene. A standard candle when burning vitiates the air in a room almost as much as an adult person. An ordinary kerosene lamp vitiates the atmosphere as much as a half-dozen persons. An ordinary single mantle burner causes as much vitiation as two or three persons.

In order to obtain a bird's-eye view of progress in light-production, the following table of relative luminous efficiencies of several light-sources is given in round numbers. These efficiencies are in terms of the most efficient (yellow-green) light.

Efficiency in per cent. Sperm-candle 0.02 Open gas-flame .04 Incandescent gas-mantle .19 Carbon filament lamp .05 Vacuum Mazda lamp 1.3 Gas-filled Mazda lamp 2 to 3 Arc-lamps 2 to 7 White light radiated by "black-body" 16 Most efficient white light 40 Firefly 95 Most efficient light (yellow-green) 100

The luminous efficiency of a light-source is distinguished from that of a lamp. The former is the ratio of the light produced to the amount of energy radiated by the light-source. The latter is the ratio of the light produced to the total amount of energy consumed by the device. In other words, the luminous efficiency of a lamp is less than that of the light-source because the consumption of energy in other parts of the lamp besides the light-source are taken into account. These additional losses are appreciable in the mechanisms of arc-lamps but are almost negligible in vacuum incandescent filament lamps. They are unknown for the firefly, so that its luminous efficiency only as a light-source can be determined. Its efficiency as a lighting-plant may be and perhaps is rather low.

VIII

MODERN GAS-LIGHTING

As has been seen, the lighting industry, as a public service, was born in London about a century ago and companies to serve the public were organized on the Continent shortly after. From this early beginning gas-light remained for a long time the only illuminant supplied by a public-service company. It has been seen that throughout the ages little advance was made in lighting until oil-lamps were improved by Argand in the eighteenth century. Candles and open-flame oil-lamps were in use when the Pyramids were built and these were common until the approach of the nineteenth century. In fact, several decades passed after the first gas-lighting was installed before this form of lighting began to displace the improved oil-lamps and candles. It was not until about 1850 that it began to invade the homes of the middle and poorer classes. During the first half of the nineteenth century the total light in an average home was less than is now obtained from a single light-source used in residences; still, the total cost of lighting a residence has decreased considerably. If the social and industrial activities of mankind are visualized for these various periods in parallel with the development of artificial lighting, a close relation is evident. Did artificial light advance merely hand in hand with science, invention, commerce, and industry, or did it illuminate the pathway?

Although gas-lighting was born in England it soon began to receive attention elsewhere. In 1815 the first attempt to provide a gas-works in America was made in Philadelphia; but progress was slow, with the result that Baltimore and New York led in the erection of gas-works. There are on record many protests against proposals which meant progress in lighting. These are amusing now, but they indicate the inertia of the people in such matters. When Bollman was projecting a plan for lighting Philadelphia by means of piped gas, a group of prominent citizens submitted a protest in 1833 which aimed to show that the consequences of the use of gas were appalling. But this protest failed and in 1835 a gas-plant was founded in Philadelphia. Thus gas-lighting, which to Sir Walter Scott was a "pestilential innovation" projected by a madman, weathered its early difficulties and grew to be a mighty industry. Continued improvements and increasing output not only altered the course of civilization by increased and adequate lighting but they reduced the cost of lighting over the span of the nineteenth century to a small fraction of its initial cost.

Think of the city of Philadelphia in 1800, with a population of about fifty thousand, dependent for its lighting wholly upon candles and oil-lamps! Washington's birthday anniversary was celebrated in 1817 with a grand ball attended by five hundred of the elite. An old report of the occasion states that the room was lighted by two thousand wax-candles. The cost of this lighting was a hundred times the cost of as much light for a similar occasion at the present time. Can one imagine the present complex activities of a city like Philadelphia with nearly two million inhabitants to exist under the lighting conditions of a century ago? To-day there are more than fifty thousand street lamps in the city--one for each inhabitant of a century ago. Of these street lamps about twenty-five thousand burn gas. This single instance is representative of gas-lighting which initiated the "light age" and nursed it through the vicissitudes of youth. The consumption of gas has grown in the United States during this time to three billion cubic feet per day. For strictly illuminating purposes in 1910 nearly one hundred billion cubic feet were used. This country has been blessed with large supplies of natural gas; but as this fails new oil-fields are constantly being discovered, so that as far as raw materials are concerned the future of gas-lighting is assured for a long time to come.

The advent of the gas-mantle is responsible for the survival of gas-lighting, because when it appeared electric lamps had already been invented. These were destined to become the formidable light-sources of the approaching century and without the gas-mantle gas-lighting would not have prospered. Auer von Welsbach was conducting a spectroscopic study of the rare-earths when he was confronted with the problem of heating these substances. He immersed cotton in solutions of these salts as a variation of the regular means for studying elements by injecting them into flames. After burning the cotton he found that he had a replica of the original fabric composed of the oxide of the metal, and this glowed brilliantly when left in the flame.

This gave him the idea of producing a mantle for illuminating purposes and in 1885 he placed such a mantle in commercial use. His first mantles were unsatisfactory, but they were improved in 1886 by the use of thoria, an oxide of thorium, in conjunction with other rare-earth oxides. His mantle was now not only stronger but it gave more light. Later he greatly improved the mantles by purifying the oxides and finally achieved his great triumph by adding a slight amount of ceria, an oxide of cerium. Welsbach is deserving of a great deal of credit for his extensive work, which overcame many difficulties and finally gave to the world a durable mantle that greatly increased the amount of light previously obtainable from gas.

The physical characteristics of a mantle depend upon the fabric and upon the rare-earths used. It must not shrink unduly when burned, and the ash should remain porous. It has been found that a mantle in which thoria is used alone is a poor light-source, but that when a small amount of ceria is added the mantle glows brilliantly. By experiment it was determined that the best proportions for the rare-earth content are one part of ceria and ninety-nine parts of thoria. Greater or less proportions of ceria decreased the light-output. The actual percentage of these oxides in the ash of the mantle is about 10 per cent., making the content of ceria about one part in one thousand.

Mantles are made by knitting cylinders of cotton or of other fiber and soaking these in a solution of the nitrates of cerium and thorium. One end of the cylinder is then sewed together with asbestos thread, which also provides the loop for supporting the mantle over the burner. After the mantle has dried in proper form, it is burned; the organic matter disappears and the nitrates are converted into oxides. After this "burning off" has been accomplished and any residual blackening is removed, the mantle is dipped into collodion, which strengthens it for shipping and handling. The collodion is a solution of gun-cotton in alcohol and ether to which an oil such as castor-oil has been added to prevent excessive shrinkage on drying.

The materials and structure of the fabric of mantles have been subjected to much study. Cotton was first used; then ramie fibers were introduced. The ramie mantle was found to possess a greater life than the cotton mantle. Later the mantles were mercerized by immersion in ammonia-water and this process yielded a stronger material. The latest development is the use of an artificial silk as the base fabric, which results in a mantle superior to previous mantles in strength, flexibility, permanence of form, and permanence of luminous property. This artificial silk mantle will permit of handling even after it has been in use for several hundred hours. This great advance appears to be due to the fact that after the artificial-silk fibers have been burned off, the fibers are solid and continuous instead of porous as in previous mantles.

The color-value of the light from mantles may be varied considerably by altering the proportions of the rare-earths. The yellowness of the light has been traced to ceria, so by varying the proportions of ceria, the color of the light may be influenced.

The inverted mantle introduced greater possibilities into gas-lighting. The light could be directed downward with ease and many units such as inverted bowls were developed. In fact, the lighting-fixtures and the lighting-effects obtainable kept pace with those of electric lighting, notwithstanding the greater difficulties encountered by the designer of gas-lighting fixtures. Many problems were encountered in designing an inverted burner operating on the Bunsen principle, but they were finally satisfactorily solved. In recent years a great deal of study has been given to the efficiency of gas-burners, with the result that a high level of development has been reached.

Several methods of electrical ignition have been evolved which in general employ the electric spark. Electrical ignition and developments of remote control have added great improvements especially to street-lighting by means of gas. Gas-valves for remote control are actuated by gas pressure and by electromagnets. In general, the gas-lighting engineers have kept pace marvelously with electric lighting, when their handicaps are considered.

Various types of burners have appeared which aimed to burn more gas in a given time under a mantle and thereby to increase the output of light. These led to the development of the pressure system in which the pressure of gas was at first several times greater than usual. The gas is fed into the mixing tube under this higher pressure in a manner which also draws in an adequate amount of air. In this way the combustion at the burner is forced beyond the point reached with the usual pressure. Ordinary gas pressure is equal to that of a few inches of water, but high-pressure systems employ pressures as great as sixty inches of water. Under this high-pressure system, mantle-burners yield as high as 500 lumens per cubic foot of gas per hour.

The fuels for gas-lighting are natural gas, carbureted water-gas, and coal-gas obtained by distilling coal, but there are different methods of producing the artificial gases. Coal-gas is produced analytically by distilling certain kinds of coal, but water-gas and producer-gas are made synthetically by the action of several constituents upon one another. Carbureted water-gas is made from fixed carbon, steam, and oil and also from steam and oil. Producer-gas is made by the action of steam or air or both upon fixed carbon. Water-gas made from steam and oil is usually limited to those places where the raw materials are readily available. The composition of a gas determines its heating and illuminating values, and constituents favorable to one are not necessarily favorable to the other. Coal-gas usually is of lower illuminating value than carbureted water-gas. It contains more hydrogen, for example, than water-gas and it is well known that hydrogen gives little light on burning.

It has been seen in a previous chapter that the distillation of gas from coal for illuminating purposes began in the latter part of the eighteenth century. From this beginning the manufacture of coal-gas has been developed to a great and complex industry. The method is essentially destructive distillation. The coal is placed in a retort and when it reaches a temperature of about 700 癬. through heating by an outside fire, the coal begins to fuse and hydrocarbon vapors begin to emanate. These are generally paraffins and olefins. As the temperature increases, these hydrocarbons begin to be affected. The chemical combinations which have long existed are broken up and there are rearrangements of the atoms of carbon and hydrogen. The actual chemical reactions become very complex and are somewhat shrouded in uncertainty. In this last stage the illuminating and heating values of the gas are determined. Usually about four hours are allowed for the complete distillation of the gaseous and liquid products from a charge of coal. Many interesting chemical problems arise in this process and the influences of temperature and time cannot be discussed within the scope of this book. Besides the coal-gas, various by-products are obtained depending upon the raw materials, upon the procedure, and upon the market.

After the coal-gas is produced it must be purified and the sulphureted hydrogen at least must be removed. One method of accomplishing this is by washing the gas with water and ammonia, which also removes some of the carbon dioxide and hydrocyanic acid. Various other undesirable constituents are removed by chemical means, depending upon the conditions. The purified gas is now delivered to the gas-holder; but, of course, all this time the pressure is governed, in order that the pressure in the mains will be maintained constant.

Much attention has been given to the enrichment of gas for illuminating purposes; that is, to produce a gas of high illuminating value from cheap fuel or by inexpensive processes. This has been done by decomposing the tar obtained during the distillation of coal and adding these gases to the coal-gas; by mixing carbureted water-gas with coal-gas; by carbureting inferior coal-gases; and by mixing oil-gas with inferior coal-gas.

Water-gas is of low illuminating value, but after it is carbureted it burns with a brilliant flame. The water-gas is made by raising the temperature of the fuel bed of hard coal or coke by forced air, which is then cut off, while steam is passed through the incandescent fuel. This yields hydrogen and carbon monoxide. To make carbureted water-gas, oil-gas is mixed with it, the latter being made by heating oil in retorts.

A great many kinds of gas are made which are determined by the requirements and the raw materials available. The amount of illuminating gas yielded by a ton of fuel, of course, varies with the method of manufacture, with the raw material, and with the use to which the fuel is to be put. The production of coal-gas per ton of coal is of the order of magnitude of 10,000 cubic feet. A typical yield by weight of a coal-gas retort is,

10,000 cubic feet of gas 17 per cent. coke 70 " " tar 5 " " ammoniacal liquid 8 " "

The coke is not pure carbon but contains the non-volatile minerals which will remain as ash when the coke is burned, just as if the original coal had been burned. On the crown of the retort used in coal-gas production, pure carbon is deposited. This is used for electric-arc carbons and for other purposes. From the tar many products are derived such as aniline dyes, benzene, carbolic acid, picric acid, napthalene, pitch, anthracene, and saccharin.

A typical analysis of the gas distilled from coal is very approximately as follows,

Hydrocarbons 40 per cent. Hydrogen 50 " " Carbon monoxide 4 " " Nitrogen 4 " " Carbon dioxide 1 " " Various other gases 1 " "

It is seen that illuminating gas is not a definite compound but a mixture of a number of gases. The proportion of these is controlled in so far as possible in order to obtain illuminating value and some of them are reduced to very small percentages because they are valueless as illuminants or even harmful. The constituents are seen to consist of light-giving hydrocarbons, of gases which yield chiefly heat, and of impurities. The chief hydrocarbons found in illuminating gas are,

ethylene C_2H_4 crotonylene C_4H_6 propylene C_3H_6 benzene C_6H_6 butylene C_4H_8 toluene C_7H_8 amylene C_5H_{10} xylene C_8H_{10} acetylene C_2H_2 methane $C H_4$ allylene C_3H_4 ethane C_2H_6

A gas which has played a prominent part in lighting is acetylene, produced by the interaction of water and calcium carbide. No other gas easily produced upon a commercial scale yields as much light, volume for volume, as acetylene. It has the great advantage of being easily prepared from raw material whose yield of gas is considerably greater for a given amount than the raw materials which are used in making other illuminating gases. The simplicity of the manufacture of acetylene from calcium carbide and water gives to this gas a great advantage in some cases. It has served for individual lighting in houses and in other places where gas or electric service was unavailable. Where space is limited it also had an advantage and was adopted to some extent on automobiles, motor-boats, ships, lighthouses, and railway cars before electric lighting was developed for these purposes.

The color of the acetylene flame is satisfactory and it is extremely brilliant compared with most flames. An interesting experiment is found in placing a spark-gap in the flame and sending a series of sparks across it. If the conditions are proper the flame will became very much brighter. When the gas issues from a proper jet under sufficient pressure, the flame is quite steady. Its luminous efficiency gives it an advantage over other open gas-flames in lighting rooms, because for the same amount of light it vitiates the air and exhausts the oxygen to a less degree than the others. Of course, in these respects the gas-mantle is superior.

The reaction which takes place when water and calcium carbide are brought together is a double decomposition and is represented by,

$$CaC_2 + H_2O = C_2H_2 + CaO$$

It will be seen that the products are acetylene gas and calcium oxide or lime. The lime, being hydroscopic and being in the presence of water or water-vapor in the acetylene generator, really becomes calcium hydroxide $Ca(OH)_2$, commonly called slaked lime. If there are impurities in the calcium carbide, it is sometimes necessary to purify the gas before it may be safely used for interior lighting.

The burners and mantles used in acetylene lighting are essentially the same as those for other gas-lighting, excepting, of course, that they are especially adapted for it in minor details.

The chief source of calcium carbide in this country is the electric furnace. Cheap electrical energy from hydro-electric developments, such as the Niagara plants, have done much to make the earth yield its elements. Aluminum is very prevalent in the soil of the earth's surface, because its oxide, alumina, is a chief constituent of ordinary clay. But the elements, aluminum and oxygen, cling tenaciously to each other and only the electric furnace with its excessively high temperatures has been able to separate them on a large commercial scale. Similarly, calcium is found in various compounds over the earth's surface. Limestone abounds widely, hence the oxide and carbonate of lime are wide-spread. But calcium clings tightly to the other elements of its compounds and it has taken the electric furnace to bring it to submission. The cheapness of calcium carbide is due to the development of cheap electric power. It is said that calcium carbide was discovered as a by-product of the electric furnace by accidentally throwing water upon the waste materials of a furnace process. The discovery of a commercial scale of manufacture of calcium carbide has been a boon to isolated lighting. Electric lighting has usurped its place on the automobile and is making inroads in country-home lighting. Doubtless, acetylene will continue to serve for many years, but its future does not appear as bright as it did many years ago.

The Pintsch gas, used to some extent in railroad passenger-cars in this country, is an oil-gas produced by the destructive distillation of petroleum or other mineral oil in retorts heated externally. The product consists chiefly of methane and heavy hydrocarbons with a small amount of hydrogen. In the early days of railways, some trains were not run after dark and those which were operated were not always lighted. At first attempts were made at lighting railway cars with compressed coal-gas, but the disadvantage of this was the large tank required. Obviously, a gas of higher illuminating-value per volume was desired where limited storage space was available, and Pintsch turned his attention to oil-gas. Gas suffers in illuminating-value upon being compressed, but oil-gas suffers only about half the loss that coal-gas does. In about 1880 Pintsch developed a method of welding cylinders and buoys which satisfied lighthouse authorities and he was enabled to furnish these filled with compressed gas. Thus the buoy was its own gas-tank. He devised lanterns which would remain lighted regardless of wind and waves and thus gained a start with his compressed-gas systems. He compressed the gas to a pressure of about one hundred and fifty pounds per square inch and was obliged to devise a reducer which would deliver the gas to the burner at about one pound per square inch. This regulator served well throughout many years of exacting service. The system began to be adopted on ships and railroads in 1880 and for many years it has served well.

Although gas-lighting has affected the activities of mankind considerably by intensifying commerce and industry and by advancing social progress, the illuminants which eventually took the lead have extended the possibilities and influences of artificial light. In the brief span of a century civilized man is almost totally independent of natural light in those fields over which he has control. What another century will bring can be predicted only from the accomplishments of the past. These indicate possibilities beyond the powers of imagination.

IX

THE ELECTRIC ARCS

Early in 1800 Volta wrote a letter to the President of the Royal Society of London announcing the epochal discovery of a device now known as the voltaic pile. This letter was published in the Transactions and it created great excitement among scientific men, who immediately began active investigations of certain electrical phenomena. Volta showed that all metals could be arranged in a series so that each one would indicate a positive electric potential when in contact with any metal following it in the series. He constructed a pile of metal disks consisting of zinc and copper alternated and separated by wet cloths. At first he believed that mere contact was sufficient, but when, later, it was shown that chemical action took place, rapid progress was made in the construction of voltaic cells. The next step after his pile was constructed was to place pairs of strips of copper and zinc in cups containing water or dilute acid. Volta received many honors for his discovery, which contributed so much to the development of electrical science and art--among them a call to Paris by Bonaparte to exhibit his electrical experiments, and to receive a medal struck in his honor.

While Volta was being showered with honors, various scientific men with great enthusiasm were entering new fields of research, among which was the heating value of electric current and particularly of electric sparks made by breaking a circuit. Late in 1800 Sir Humphrey Davy was the first to use charcoal for the sparking points. In a lecture before the Royal Society in the following year he described and demonstrated that the "spark" passing between two pieces of charcoal was larger and more brilliant than between brass spheres. Apparently, he was producing a feeble arc, rather than a pure spark. In the years which immediately followed many scientific men in England, France, and Germany were publishing the results of their studies of electrical phenomena bordering upon the arc.

By subscription among the members of the Royal Society, a voltaic battery of two thousand cells was obtained and in 1808 Davy exhibited the electric arc on a large scale. It is difficult to judge from the reports of these early investigations who was the first to recognize the difference between the spark and the arc. Certainly the descriptions indicate that the simple spark was not being experimented with, but the source of electric current available at that time was of such high resistance that only feeble arcs could have been produced. In 1809 Davy demonstrated publicly an arc obtained by a current from a Volta pile of one thousand plates. This he described as "a most brilliant flame, of from half an inch to one and a quarter inches in length."

In the library of the Royal Society, Davy's notes made during the years of 1805 and 1812 are available in two large volumes. These were arranged and paged by Faraday, who was destined to contribute greatly to the future development of the science and art of electricity. In one of these volumes is found an account of a lecture-experiment by Davy which certainly is a description of the electric arc. An extract of this account is as follows:

The spark [presumably the arc], the light of which was so intense as to resemble that of the sun, ... produced a discharge through heated air nearly three inches in length, and of a dazzling splendor. Several bodies which had not been fused before were fused by this flame.... Charcoal was made to evaporate, and plumbago appeared to fuse in vacuo. Charcoal was ignited to intense whiteness by it in oxymuriatic acid, and volatilized by it, but without being decomposed.

From a consideration of his source of electricity, a voltaic pile of two thousand plates, it is certain that this could not have been an electric spark. Later in his notes Davy continued:

...the charcoal became ignited to whitness, and by withdrawing the points from each other, a constant discharge took place through the heated air, in a space at least equal to four inches, producing a most brilliant ascending arch of light, broad and conical in form in the middle.

This is surely a description of the electric arc. Apparently the electrodes were in a horizontal position and the arc therefore was horizontal. Owing to the rise of the heated air, the arc tended to rise in the form of an arch. From this appearance the term "arc" evolved and Davy himself in 1820 definitely named the electric flame, the "arc." This name was continued in use even after the two carbons were arranged in a vertical co-axial position and the arc no more "arched." An interesting scientific event of 1820 was the discovery by Arago and by Davy independently that the arc could be deflected by a magnet and that it was similar to a wire carrying current in that there was a magnetic field around it. This has been taken advantage of in certain modern arc-lamps in which inclined carbons are used. In these arcs a magnet keeps the arc in place, for without the magnet the arc would tend to climb up the carbons and go out.

In 1838 Gassiot made the discovery that the temperature of the positive electrode of an electric arc is much greater than that of the negative electrode. This is explained in electronic theory by the bombardment of the positive electrode by negative electrons or corpuscles of electricity. This temperature-difference was later taken into account in designing direct-current arc-lamps, for inasmuch as most of the light from an ordinary arc is emitted by the end of the positive electrode, this was placed above the negative electrode. In this manner most of the light from the arc is directed downward where desired. In the few instances in modern times where the ordinary direct-current arc has been used for indirect lighting, in which case the arc is above an inverted shade, the positive carbon is placed below the negative one. Gassiot first proved that the positive electrode is hotter than the negative one by striking an arc between the ends of two horizontal wires of the same substance and diameter. After the arc operated for some time, the positive wire was melted for such a distance that it bent downward, but the negative remained quite straight.

Charcoal was used for the electrodes in all the early experiments, but owing to the intense heat of the arc, it burned away rapidly. A progressive step was made in 1843 when electrodes were first made by Foucault from the carbon deposited in retorts in which coal was distilled in the production of coal-gas. However, charcoal, owing to its soft porous character, gives a longer arc and a larger flame. In 1877 the "cored" carbons were introduced. These consist of hard molded carbon rods in which there is a core of soft carbon. In these are combined the advantages of charcoal and hard carbon and the core in burning away more rapidly has a tendency to hold the arc in the center. Modern carbons for ordinary arc-lamps are generally made of a mixture of retort-carbon, soot, and coal-tar. This paste is forced through dies and the carbons are baked at a fairly high temperature. A variation in the hardness of the carbons may be obtained as the requirements demand by varying the proportions of soot and retort-carbon. Cored carbons are made by inserting a small rod in the center of the die and the carbons are formed with a hollow core. This may be filled with a softer carbon.

If two carbons connected to a source of electric current are brought together, the circuit is completed and a current flows. If the two carbons are now slightly separated, an arc will be formed. As the arc burns the carbons waste away and in the case of direct current, the positive decreases in length more rapidly than the negative one. This is due largely to the extremely high temperature of the positive tip, where the carbon fairly boils. A crater is formed at the positive tip and this is always characteristic of the positive carbon of the ordinary arc, although it becomes more shallow as the arc-length is increased. The negative tip has a bright spot to which one end of the arc is attached. By wasting away, the length of the arc increases and likewise its resistance, until finally insufficient current will pass to maintain the arc. It then goes out and to start it the carbons must be brought together and separated. The mechanisms of modern arc-lamps perform these functions automatically by the ingenious use of electromagnets.

The interior of the arc is of a violet color and the exterior is a greenish yellow. The white-hot spot on the negative tip is generally surrounded by a fringe of agitated globules which consist of tar and other ingredients of carbons. Often material is deposited from the positive crater upon the negative tip and these accretions may build up a rounded tip. This deposit sometimes interferes with the proper formation of the arc and also with the light from the arc. It is often responsible for the hissing noise, although this hissing occurs with any length of arc when the current is sufficiently increased. The hissing seems to be due to the crater enlarging under excessive current until it passes the confines of the cross-section of the carbon. It thus tends to run up the side, where it comes in contact with oxygen of the air. In this manner the carbon is directly burned instead of being vaporized, as it is when the hot crater is small and is protected from the air by the arc itself. The temperature of the positive crater is in the neighborhood of 6000?to 7000 癥. The brightness of the arc under pressure is the greatest produced by artificial means and is very intense. By putting the arc under high pressure, the brightness of the sun may be attained. The temperature of the hottest spot on the negative tip is about a thousand degrees below that of the positive.

No great demand arose for arc-lamps until the development of the Gramme dynamo in 1870, which provided a practicable source of electric current. In

1876 Jablochkov invented his famous "electric candle" consisting of two rods of carbon placed side by side but separated by insulating material. In this country Brush was the pioneer in the development of open arc-lamps. In 1877 he invented an arc-lamp and an efficient form of dynamo to supply the electrical energy. The first arc-lamps were ordinary direct-current open arcs and the carbons were made from high-grade coke, lampblack, and syrup. The upper positive carbon in these lamps is consumed at a rate of one to two inches per hour. Inasmuch as about 85 per cent. of the total light is emitted by the upper (positive) carbon and most of this from the crater, the lower carbon is made as small as possible in order not to obstruct any more light than necessary. The positive carbon of the open arc is often cored and the negative is a smaller one of solid carbon. This combination operates quite satisfactorily, but sometimes solid carbons are used outdoors. The voltage across the arc is about 50 volts.

In 1846 Staite discovered that the carbons of an arc enclosed in a glass vessel into which the air was not freely admitted were consumed less rapidly than when the arc operated in the open air. After the appearance of the dynamo, when increased attention was given to the development of arc-lamps, this principle of enclosing the arcs was again considered. The early attempts in about 1880 were unsuccessful because low voltages were used and it was not until the discovery was made that the negative tip builds up considerably for voltages under 65 volts, that higher voltages were employed. In 1893 marked improvements were consummated and Jandus brought out a successful enclosed arc operating at 80 volts. Marks contributed largely to the success of the enclosed arc by showing that a small current and a high voltage of 80 to 85 volts were the requisites for a satisfactory enclosed arc.

The principle of the enclosed arc is simple. A closely fitting glass globe surrounds the arc, the fit being as close as the feeding of the carbons will permit. When the arc is struck the oxygen is rapidly consumed and the heated gases and the enclosure check the supply of fresh air. The result is that the carbons are consumed about one tenth as rapidly as in the open arc. There is no crater formed on the positive tip and the arc wanders considerably. The efficiency of the enclosed arc as a light-producer is lower than that of the open arc, but it found favor because of its slow rate of consumption of the carbons and consequent decreased attention necessary. This arc operates a hundred hours or more without trimming, and will therefore operate a week or more in street-lighting without attention. When it is considered that open arcs for all-night burning were supplied with two pairs of carbons, the second set going into use automatically when the first were consumed, the value of the enclosed arc is apparent. However, the open arc has served well and has given way to greater improvements. It is rapidly disappearing from use.

The alternating-current arc-lamp was developed after the appearance of the direct-current open-arc and has been widely used. It has no positive or negative carbons, for the alternating current is reversing in direction usually at the rate of 120 times per second; that is, it passes through 60 complete cycles during each second. No marked craters form on the tips and the two carbons are consumed at about the same rate. The average temperature of the carbon tips is lower than that of the positive tip of a direct-current arc, with the result that the luminous efficiency is lower. These arcs have been made of both the open and enclosed type. They are characterized by a humming noise due to the effect of alternating current upon the mechanism and also upon the air near the arc. This humming sound is quite different from the occasional hissing of a direct-current arc. When soft carbons are used, the arc is larger and apparently this mass of vapor reduces the humming considerably. The humming is not very apparent for the enclosed alternating-current arc. The alternating arc can easily be detected by closely observing moving objects. If a pencil or coin be moved rapidly, a number of images appear which are due to the pulsating character of the light. At each reversal of the current, the current reaches zero value and the arc is virtually extinguished. Therefore, there is a maximum brightness midway between the reversals.

Various types of all these arcs have been developed to meet the different requirements of ordinary lighting and to adapt this method of light-production to the needs of projection, stage-equipment, lighthouses, search-lights, and other applications.

Up to this point the ordinary carbon arc has been considered and it has been seen that most of the light is emitted by the glowing end of the positive carbon. In fact, the light from the arc itself is negligible. A logical step in the development of the arc-lamp was to introduce salts in order to obtain a luminous flame. This possibility as applied to ordinary gas-flames had been known for years and it is surprising that it had not been early applied to carbons. Apparently Bremer in 1898 was the first to introduce fluorides of calcium, barium, and strontium. The salts deflagrate and a luminous flame envelops the ordinary feeble arc-flame. From these arcs most of the light is emitted by the arc itself, hence the name "flame-arcs."

By the introduction of metallic salts into the carbons the possibilities of the arc-lamp were greatly extended. The luminous output of such lamps is much greater than that of an ordinary carbon arc using the same amount of electrical energy. Furthermore, the color or spectral character of the light may be varied through a wide range by the use of various salts. For example, if carbons are impregnated with calcium fluoride, the arc-flame when examined by means of a spectroscope will be seen to contain the characteristic spectrum of calcium, namely, some green, orange, and red rays. These combine to give to this arc a very yellow color. As explained in a previous chapter, the salts for this purpose may be wisely chosen from a knowledge of their fundamental or characteristic flame-spectra.

These lamps have been developed to meet a variety of needs and their luminous efficiencies range from 20 to 40 lumens per watt, being several times that of the ordinary carbon open-arc. The red flame-arc owes its color chiefly to strontium, whose characteristic visible spectrum consists chiefly of red and yellow rays. Barium gives to the arc a fairly white color. The yellow and so-called white flame-arcs have been most commonly used. Flame-arcs have been produced which are close to daylight in color, and powerful blue-white flame-arcs have satisfied the needs of various chemical industries and photographic processes. These arcs are generally operated in a space where the air-supply is restricted similar to the enclosed-arc principle. Inasmuch as poisonous fumes are emitted in large quantities from some flame-arcs, they are not used indoors without rather generous ventilation. In fact, the flame-arcs are such powerful light-sources that they are almost entirely used outdoors or in very large interiors especially of the type of open factory buildings. They are made for both direct and alternating current and the mechanisms have been of several types. The electrodes are consumed rather rapidly so they are made as long as possible. In one type of arc, the carbons are both fed downward, their lower ends forming a narrow V with the arc-flame between their tips. Under these conditions the arc tends to travel vertically and finally to "stretch" itself to extinction. However, the arc is kept in place by means of a magnet above it which repels the arc and holds it at the ends of the carbons.

The chief objection to the early flame-arcs was the necessity for frequent renewal of the carbons. This was overcome to a large extent in the Jandus regenerative lamp in which the arc operates in a glass enclosure surrounded by an opal globe. However, in addition to the inner glass enclosure, two cooling chambers of metal are attached to it. Air enters at the bottom and the fumes from the arc pass upward and into the cooling chambers, where the solid products are deposited. The air on returning to the bottom is thus relieved of these solids and the inner glass enclosure remains fairly clean. The lower carbon is impregnated with salts for producing the luminous flame and the upper carbon is cored. The life of the electrodes is about seventy-five hours.

The next step was the introduction of the so-called "luminous-arc" which is a "flame-arc" with entirely different electrodes. The lower (negative) electrode consists of an iron tube packed chiefly with magnetite (an iron oxide) and titanium oxide in the approximate proportions of three to one respectively. The magnetite is a conductor of electricity which is easily vaporized. The arc-flame is large and the titanium gives it a high brilliancy. The positive electrode, usually the upper one, is a short, thick, solid cylinder of copper, which is consumed very slowly. This lamp, known as the magnetite-arc, has a luminous efficiency of about 20 lumens per watt with a clear glass globe.

The mechanisms which strike the arc and feed the carbons are ingenious devices of many designs depending upon the kind of arc and upon the character of the electric circuit to which it is connected. Late developments in electric incandescent filament lamps have usurped some of the fields in which the arc-lamp reigned supreme for years and its future does not appear as bright now as it did ten years ago. High-intensity arcs have been devised with small carbons for special purposes and considered as a whole a great amount of ingenuity has been expended in the development of arc-lamps. There will be a continued demand for arc-lamps, for scientific developments are opening new fields for them. Their value in photo-engraving, in the moving-picture production studios, in moving-picture projection, and in certain aspects of stage-lighting is firmly established, and it appears that they will find application in certain chemical industries because the arc is a powerful source of radiant energy which is very active in its effects upon chemical reactions.

The luminous efficiencies of arc-lamps depend upon so many conditions that it is difficult to present a concise comparison; however, the following may suffice to show the ranges of luminous output per watt under actual conditions of usage. These efficiencies, of course, are less than the efficiencies of the arc alone, because the losses in the mechanism, globes, etc., are included.

Lumens per watt Open carbon arc 4 to 8 Enclosed carbon arc 3 to 7 Enclosed flame-arc (yellow or white) 15 to 25 Luminous arc 10 to 25

Another lamp differing widely in appearance from the preceding arcs may be described here because it is known as the mercury-arc. In this lamp mercury is confined in a transparent tube and an arc is started by making and breaking a mercury connection between the two electrodes. The arc may be maintained of a length of several feet. Perhaps the first mercury-arc was produced in 1860 by Way, who permitted a fine jet of mercury to fall from a reservoir into a vessel, the reservoir and receiver being connected to the poles of a battery. The electric current scattered the jet and between the drops arcs were formed. He exhibited this novel light-source on the mast of a yacht and it received great attention. Later, various investigators experimented on the production of a mercury-arc and the first successful ones were made in the form of an inverted U-tube with the ends filled with mercury and the remainder of the tube exhausted.

Cooper Hewitt was a successful pioneer in the production of practicable mercury-arcs. He made them chiefly in the form of straight tubes of glass up to several feet in length, with enlarged ends to facilitate cooling. The tubes are inclined so that the mercury vapor which condenses will run back into the enlarged end, where a pool of mercury forms the negative electrode. The arc may be started by tilting the tube so that a mercury thread runs down the side and connects with the positive electrode of iron. The heat of the arc volatilizes the mercury so that an arc of considerable length is maintained. The tilting is done by electromagnets. Starting has also been accomplished by means of a heating coil and also by an electric spark. The lamps are stabilized by resistance and inductance coils.

One of the defects of the light emitted by the incandescent vapor of mercury is its paucity of spectral colors. Its visible spectrum consists chiefly of violet, blue, green, and yellow rays. It emits virtually no red rays, and, therefore, red objects appear devoid of red. The human face appears ghastly under this light and it distorts colors in general. However, it possesses the advantages of high efficiency, of reasonably low brightness, of high actinic value, and of revealing detail clearly. Various attempts have been made to improve the color of the light by adding red rays. Reflectors of a fluorescent red dye have been used with some success, but such a method reduces the luminous efficiency of the lamp considerably. The dye fluoresces red under the illumination of ultra-violet, violet, and blue rays; that is, it has the property of converting radiation of these wave-lengths into radiant energy of longer wave-lengths. By the use of electric incandescent filament lamps in conjunction with mercury-arcs, a fairly satisfactory light is obtained. Many experiments have been made by adding other substances to the mercury, such as zinc, with the hope that the spectrum of the other substance would compensate the defects in the mercury spectrum. However no success has been reached in this direction.

By the use of a quartz tube which can withstand a much higher temperature than glass, the current density can be greatly increased. Thus a small quartz tube of incandescent mercury vapor will emit as much light as a long glass tube. The quartz mercury-arc produces a light which is almost white, but the actual spectrum is very different from that of white sunlight. Although some red rays are emitted by the quartz arc, its spectrum is essentially the same as that of the glass-tube arc. Quartz transmits ultra-violet radiation, which is harmful to the eyes, and inasmuch as the mercury vapor emits such rays, a glass globe should be used to enclose the quartz tube when the lamp is used for ordinary lighting purposes.

It is fortunate that such radically different kinds of light-sources are available, for in the complex activities of the present time all are in demand. The quartz mercury-arc finds many isolated uses, owing to its wealth of ultra-violet radiation. It is valuable as a source of ultra-violet for exciting phosphorescence, for examining the transmission of glasses for this radiation, for sterilizing water, for medical purposes, and for photography.

X

THE ELECTRIC INCANDESCENT FILAMENT LAMPS

Prior to 1800 electricity was chiefly a plaything for men of scientific tendencies and it was not until Volta invented the electric pile or battery that certain scientific men gave their entire attention to the study of electricity. Volta was not merely an inventor, for he was one of the greatest scientists of his period, endowed with an imagination which marked him as a genius in creative work. By contributing the electric battery, he added the greatest impetus to research in electrical science that it has ever received. As has already been shown, there began a period of enthusiastic research in the general field of heating effects of electric current. The electric arc was born in the cradle of this enthusiasm, and in the heating of metals by electricity the future incandescent lamp had its beginning.

Between the years 1841 and 1848 several inventors attempted to make light-sources by heating metals. These crude lamps were operated by means of Grove and Bunsen electric cells, but no practicable incandescent filament lamps were brought out until the development of the electric dynamo supplied an adequate source of electric current. As electrical science progressed through the continued efforts of scientific men, it finally became evident that an adequate supply of electric current could be obtained by mechanical means; that is, by rotating conductors in such a manner that current would be generated within them as they cut through a magnetic field. Even the pioneer inventors of electric lamps made great contributions to electrical practice by developing the dynamo. Brush developed a satisfactory dynamo coincidental with his invention of the arc-lamp, and in a similar manner, Edison made a great contribution to electrical practice in devising means of generating and distributing electricity for the purpose of serving his filament lamp.

[Illustration: DIRECT CURRENT ARC

Most of the light being emitted by the positive (upper) electrode]

Thousands of lamps are burned out for the sake of making improvements.

The electrical energy used is equivalent to that consumed by a city of 30,000 inhabitants]

Edison in 1878 attacked the problem of producing light from a wire or filament heated electrically. He used platinum wire in his first experiments, but its volatility and low melting-point (3200.) limited the success of the lamps. Carbon with its extremely high melting-point had long attracted attention and in 1879 Edison produced a carbon filament by carbonizing a strip of paper. He sealed this in a vessel of glass from which the air was exhausted and the electric current was led to the filament through platinum wires sealed in the glass. Platinum was used because its expansion and contraction is about the same as glass. Incidentally, many improvements were made in incandescent lamps and thirty years passed before a material was found to replace the platinum leading-in wires. The cost of platinum steadily increased and finally in the present century a substitute was made by the use of two metals whose combined expansion was the same as that of platinum or glass. In 1879 and 1880 Edison had succeeded in overcoming the many difficulties sufficiently to give to the world a practicable incandescent filament lamp. About this time Swan and Stearn in England had also produced a successful lamp.

In Edison's early experiments with filaments he used platinum wire coated with carbon but without much success. He also made thin rods of a mixture of finely divided metals such as platinum and iridium mixed with such oxides as magnesia, zirconia, and lime. He even coiled platinum wire around a piece of one of these oxides, with the aim of obtaining light from the wire and from the heated oxide. However, these experiments served little purpose besides indicating that the filament was best if it consisted solely of carbon and that it should be contained in an evacuated vessel.

One of the chief difficulties was to make the carbon filaments. Some of the pioneers, such as Sawyer and Mann, attempted to cut these from a piece of carbon. However, Edison and also Swan turned their attention to forming them by carbonizing a fiber of organic matter. Filaments cut from paper and threads of cotton and silk were carbonized for this purpose. Edison scoured the earth for better materials. He tried a fibrous grass from South America and various kinds of bamboo from other parts of the world. Thin filaments of split bamboo eventually proved the best material up to that time. He made many lamps containing filaments of this material, and even until 1910 bamboo was used to some extent in certain lamps.

Of these early days, Edison said:

It occurred to me that perhaps a filament of carbon could be made to stand in sealed glass vessels, or bulbs, which we were using, exhausted to a high vacuum. Separate lamps were made in this way independent of the air-pump, and, in October, 1879, we made lamps of paper carbon, and with carbons of common sewing thread, placed in a receiver or bulb made entirely of glass, with the leading-in wires sealed in by fusion. The whole thing was exhausted by the Sprengel pump to nearly one-millionth of an atmosphere. The filaments of carbon, although naturally quite fragile owing to their length and small mass, had a smaller radiating surface and higher resistance than we had dared hope. We had virtually reached the position and condition where the carbons were stable. In other words, the incandescent lamp as we still know it to-day [1904], in essentially all its particulars unchanged, had been born.

After Edison's later success with bamboo, Swan invented a process of squirting filaments of nitrocellulose into a coagulating liquid, after which they are carbonized. Very fine uniform filaments can be made by this process and although improvements have been made from time to time, this method has been employed ever since its invention. In these later years cotton is dissolved in a suitable solvent such as a solution of zinc chloride and this material is forced through a small diamond die. This thread when hardened appears similar to cat-gut. It is cut into proper lengths and bent upon a form. It is then immersed in plumbago and heated to a high temperature in order to destroy the organic matter. A carbon filament is the result. From this point to the finished lamp many operations are performed, but a discussion of these would lead far afield. The production of a high vacuum is one of the most important processes and manufacturers of incandescent lamps have mastered the art perhaps more thoroughly than any other manufacturers. At least, their experience in this field made it possible for them to produce quickly and on a large scale such devices as X-ray tubes during the recent war.

During the early years of incandescent lamps, improvements were made from time to time which increased the life and the luminous efficiency of the carbon filaments, but it was not until 1906 that any radical improvement was achieved. In that year in this country a process was devised whereby the carbon filament was made more compact. In fact, from its appearance it received the name "metallized filament." These carbon filaments are prepared in the same manner as the earlier ones but are finally "treated" by heating in an atmosphere of hydrocarbons such as coal-gas. The filament is heated by electric current and the heat breaks down the hydrocarbons, with the result that carbon is deposited upon the filament. This "treated" filament has a coating of hard carbon and its electrical resistance is greater than that of the untreated filament.

The luminous efficiency of a carbon filament is a function of its temperature and it increases very rapidly with increasing temperature. For this reason it is a constant aim to reach high filament temperatures. Of all the materials used in filaments up to the present time, carbon possesses the highest melting-point (perhaps as high as 7000.), but the carbon filament as operated in practice has a lower efficiency than any other filament. This is because the highest temperature at which it can be operated and still have a reasonable life is much lower than that of metallic filaments. The incandescent carbon in the evacuated bulb sublimes or volatilizes and deposits upon the bulb. This decreases the size of the filament eventually to the breaking-point and the blackening of the bulb decreases the output of light. The treated filament was found to be a harder form of carbon that did not volatilize as rapidly as the untreated filament. It immediately became possible to operate it at a higher temperature with a resulting increase of luminous efficiency. This "graphitized" carbon filament lamp became known as the gem lamp in this country and many persons have wondered over the word "gem." The first two letters stand for "General Electric" and the last for "metallized." This lamp was welcomed with enthusiasm in its day, but the day for carbon filaments has passed. The advent of incandescent lamps of higher efficiency has made it uneconomical to use carbon lamps for general lighting purposes. Although the treated carbon filament was a great improvement, its reign was cut short by the appearance of metal filaments.

In 1803 a new element was discovered and named tantalum. It is a dark, lustrous, hard metal. Pure tantalum is harder than steel; it may be drawn into fine wire; and its melting-point is very high (about 5100.). It is seen to possess properties desirable for filaments, but for some reason it did not attract attention for a long time. A century elapsed after its discovery before von

Bolton produced the first tantalum filament lamp. Owing to the low electrical resistance of tantalum, a filament in order to operate satisfactorily on a standard voltage must be long and thin. This necessitates storing away a considerable length of wire in the bulb without permitting the loops to come into contact with each other. After the filaments have been in operation for a few hundred hours they become brittle and faults develop. When examined under a microscope, parts of the filament operated on alternating current appear to be offset. The explanation of this defect goes deeply into crystalline structure. The tantalum filament was quickly followed by osmium and by tungsten in this country.

The osmium filament appeared in 1905 and its invention is due to Welsbach, who had produced the marvelous gas-mantle. Owing to its extreme brittleness, osmium was finely divided and made into a paste of organic material. The filaments were squirted through dies and, after being formed and dried, they were heated to a high temperature. The organic matter disappeared and the fine metallic particles were sintered. This made a very brittle lamp, but its high efficiency served to introduce it.

In 1870 when Scheele discovered a new element, known in this country as tungsten, no one realized that it was to revolutionize artificial lighting and to alter the course of some of the byways of civilization. This metal--which is known as "wolfram" in Germany, and to some extent in English-speaking countries--is one of the heaviest of elements, having a specific gravity of 19.1. It is 50 per cent. heavier than mercury and nearly twice as heavy as lead. It was early used in German silver to the extent of 1 or 2 per cent. to make platinoid, an alloy possessing a high resistance which varies only slightly as the temperature changes. This made an excellent material for electrical resistors. The melting-point of tungsten is about 5350., which makes it desirable for filaments, but it was very brittle as prepared in the early experiments. It unites very readily with oxygen and with carbon at high temperatures.

The first tungsten lamps appeared on the market in 1906, but these contained fragile filaments made by the squirting process. When the squirted filament of tungsten powder and organic matter was heated in an atmosphere of steam and hydrogen to remove the binding material, a brittle filament of tungsten was obtained. The first lamps were costly and fragile.

After years of organized research tungsten is now drawn into the finest wires, possessing a tensile strength perhaps greater than any other material. Filaments are now made into many shapes and the greatest strides in artificial lighting have been due to scientific research on a huge scale.

The achievements which combined to perfect the tungsten lamp to the point where it has become the mainstay of electric lighting are not attached to names in the Hall of Fame. Organization of scientific research in the industrial laboratories is such that often many persons contribute to the development of an improvement. Furthermore, time is usually required for a full perspective of applications of scientific knowledge. In the early days organized research was not practised and the great developments of those days were the works of individuals. To-day, even in pure science, some of the greatest contributions are made by industrial laboratories; but sometimes these do not become known to the public for many years. The whole scheme of scientific development has changed materially. For example, the story of the development of ductile tungsten, which has revolutionized lighting, is complex and more or less shrouded in secrecy at the present time. Many men have contributed toward this accomplishment and the public at the present time knows little more than the fact that tungsten filaments, which were brittle yesterday, are now made of ductile tungsten wire drawn into the finest filaments.

The earlier tungsten filaments were made by three rival processes. By the first, a deposit of tungsten was "flashed" on a fine carbon filament, the latter being eliminated finally by heating in an atmosphere of hydrogen and water-vapor. By the second, colloidal tungsten was produced by operating an arc between tungsten electrodes under water. The finely divided tungsten was gathered, partially dried, and squirted through dies to form filaments. These were then sintered. The third was the "paste" process already described. These methods produced fragile filaments, but their luminous efficiency was higher than that of previous ones. However, in this country ductile tungsten was soon on its way. An ingot of tungsten is subjected to vigorous swaging until it takes the form of a rod. This is finally drawn into wire.

Much of this development work was done by the laboratories of the General Electric Company and they were destined to contribute another great improvement. The blackening of the lamp bulbs was due to the evaporation of tungsten from the filament. All filaments up to this time had been confined in evacuated bulbs and the low pressure facilitates evaporation, as is well known. It had long been known that an inert gas in the bulb would reduce the evaporation and remedy other defects; however, under these conditions, there would be a considerable loss of energy through conduction of heat by the gases. In the vacuum lamp nearly all the electrical energy is converted into radiant energy, which is emitted by the filament and any dissipation of heat is an energy loss. A high vacuum was one of the chief aims up to this time, but a radical departure was pending.

If an ordinary tungsten-lamp bulb be filled with an inert gas such as nitrogen, the filament may be operated at a very much higher temperature without any more deterioration than takes place in a vacuum at a lower temperature. This gives a more efficient light but a less efficient lamp. The greater output of light is compensated by losses by conduction of heat through the gas. In other words, a great deal more energy is required by the filament in order to remain at a given temperature in a gas than in a vacuum. However, elaborate studies of the dependence of heat-losses upon the size and shape of the filament and of the physics of conduction from a solid to a gas, established the foundation for the gas-filled tungsten lamp. The knowledge gained in these investigations indicated that a thicker filament lost a relatively less percentage of energy by conduction than a thin one for equal amounts of emitted light. However, a practical filament must have sufficient resistance to be used safely on lighting circuits already established and, therefore, the large diameter and high resistance were obtained by making a helical coil of a fine wire. In fact, the gas-filled tungsten lamp may be thought of as an ordinary lamp with its long filament made into a short helical coil and the bulb filled with nitrogen or argon gas.

This development was not accidental and from a scientific point of view it is not spectacular. It did not mark a new discovery in the same sense as the discovery of X-rays. However, it is an excellent example of the great rewards which come to systematic, thorough study of rather commonplace physical laws in respect to a given condition. Such achievements are being duplicated in various lines in the laboratories of the industries. Scientific research is no longer monopolized by educational institutions. The most elaborate and best-equipped laboratories are to be found in the industries sometimes surrounded by the smoke and noise and vigorous activity which indicate that achievements of the laboratory are on their way to mankind. The smoke-laden industrial district, pulsating with life, is the proud exhibit of the present civilization. It is the creation of those who discover, organize, and apply scientific facts. But how many appreciate the debt that mankind owes not only to the individual who dedicates his life to science but to the far-sighted manufacturer who risks his money in organized quest of new benefits for mankind? A glimpse into a vast organization of research, which, for example, has been mainly responsible for the progress of the incandescent lamp would alter the attitude of many persons toward science and toward the large industrial companies.

The progress in the development of electric incandescent lamps is shown in the following table, where the dates and values are more or less approximate. It should be understood that from 1880 to the present time there has been a steady progress, which occasionally has been greatly augmented by sudden steps.

APPROXIMATE VALUES

Lumens per Date Filament Temperature watt 1880 Carbon 3300. 3.0 1906 Carbon (graphitized) 3400 4.5 1905 Tantalum 3550 6.5 1905 Osmium 3600 7.5 1906 Tungsten (vacuum) 3700 8.0 1914 Tungsten (gas-filled) up to 5300. 10 to 25

Throughout the development of incandescent filament lamps many ingenious experiments were made which resulted usually in light-sources of scientific interest but not of practical value. One of the latest is the tungsten arc in an inert gas. By means of a heating coil, a small arc is started between two electrodes consisting of tungsten, but this as yet has not been shown to be practicable.

Another type of filament lamp was developed by Nernst in 1897. It was an ingenious application of the peculiar properties of rare-earth oxides. His first lamp consisted essentially of a slender rod of magnesia. This substance does not conduct electricity at ordinary temperatures, but when heated to incandescence it becomes conducting. Upon sufficient heating of this filament by external means while a proper voltage is impressed upon it, the electric current passes through it and thereafter this current will maintain its temperature. Thus such a filament becomes a conductor and will continue to glow brilliantly by virtue of the electrical energy which it converts into heat. Later lamps consisted of "glowers" about one inch long made from a mixture of zirconia and yttria, and finally a mixture of ceria, thoria, and zirconia was used. The glower is heated initially by a coil of platinum wire located near it but not in contact with it. Owing to the fact that this glower decreases rapidly in resistance as its temperature is increased, it is necessary to place in series with it a substance which increases in resistance with increasing current. This is called a "ballasting resistance" and is usually an iron wire in a glass bulb containing hydrogen. The heater is cut out by an electromagnet when the glower goes into operation. This lamp is a marvel of ingenuity and when at its zenith it was installed to a considerable extent. Its light is considerably whiter than that of the carbon filament lamps. However, its doom was sounded when metallic filament lamps appeared.

An interesting filament was developed by Parker and Clark by using as a core a small filament of carbon. This flashed in an atmosphere containing a vapor of a compound of silicon, became coated with silicon. This filament was of high specific resistance and appeared to have promise. It has not been introduced commercially and doubtless it cannot compete with the latest tungsten lamps.

Electric incandescent lamps are the present mainstay of electric illumination and, it might be stated, of progress in lighting. Wonderful achievements have been accomplished in other modes of lighting and the foregoing statement is not meant to depreciate those achievements. However, the incandescent filament lamp has many inherent advantages. The light-source is enclosed in an air-tight bulb which makes for a safe, convenient lamp. The filament is capable of subdivision, with the result that such lamps vary from the minutest spark of the smallest miniature lamp to the enormous output of the largest gas-filled tungsten lamp. The outputs of these are respectively a fraction of a lumen and twenty-five thousand lumens; that is, the luminous intensity varies from an equivalent of a small fraction of a standard candle to a single light-source emitting light equivalent to two thousand standard candles.

Statistics are cold facts and are usually uninteresting in a volume of this character, but they tell a story in a concise manner. The development of the modern incandescent lamp has increased the intensity of light available with a great decrease in cost, and this progressive development is shown easily by tables. For example, since the advent of the tungsten lamp the average candle-power and luminous efficiency of all the lamps sold in this country has steadily increased, while the average wattages of the lamps have remained virtually stationary.

AVERAGE CANDLE-POWER, WATTS, AND EFFICIENCY OF ALL THE LAMPS SOLD IN THIS COUNTRY

Year	Candle-power	Watts	Lumens per watt
1907	18.0	53	3.33
1908	19.0	53	3.52
1909	21.0	52	3.96
1910	23.0	51	4.42
1911	25.0	51	4.82
1912	26.0	49	5.20
1913	29.4	47	6.13
1914	38.2	48	7.80
1915	42.2	47	8.74
1916	45.8	49	9.60
1917	48.7	51	10.56

It will be noted that the luminous intensity of incandescent filament lamps has steadily increased since the carbon lamp was superseded, and that in a period of ten years of organized research behind the tungsten lamp the luminous efficiency (lumens per watt) has trebled. In other words, everything else remaining unchanged, the cost of light in ten years was reduced to one third. But the reduction in cost has been more than this, as will be shown later. During the same span of years the percentage of carbon filament lamps of the total filament lamps sold decreased from 100 per cent. in 1907 to 13 per cent. in 1917. At the same time the percentage of tungsten (Mazda) lamps increased from virtually zero in 1907 to about 87 per cent. in 1917. The tantalum lamp had no opportunity to become established, because the tungsten lamp followed its appearance very closely. In 1910 the sales of the former reached their highest mark, which was only 3.5 per cent. of all the lamps sold in the United States. From a lowly beginning the number of incandescent filament lamps sold for use in this country has grown rapidly, reaching nearly two hundred million in 1919.

XI

THE LIGHT OF THE FUTURE

In viewing the development of artificial light and its manifold effects upon the activities of mankind, it is natural to look into the future. Jules Verne possessed the advantage of being able to write into fiction what his riotous imagination dictated, and so much of what he pictured has come true that his success tempts one to do likewise in prophesying the future of lighting. Surely a forecast based alone upon the past achievements and the present indications will fall short of the actual realizations of the future! If the imagination is permitted to view the future without restrictions, many apparently far-fetched schemes may be devised. It may be possible to turn to nature's supply of daylight and to place some of it in storage for night use. One millionth part of daylight released as desired at night would illuminate sufficiently all of man's nocturnal activities. The fictionist need not heed the scientist's inquiry as to how this daylight would be bottled. Instead of giving time to such inquiries he would pass on to another scheme, whereby earth would be belted with optical devices so that day could never leave. When the sun was shining in China its light would be gathered on a large scale and sent eastward and westward in these great optical "pipe-lines" to the regions of darkness, thus banishing night forever. The writer of fiction need not bother with a consideration of the economic situation which would demand such efforts. This line of conjecture is interesting, for it may suggest possibilities toward which the present trend of artificial lighting does not point; however, the author is constrained to treat the future of light-production on a somewhat more conservative basis.

At the present time the light-source of chief interest in electric lighting is the incandescent filament lamp; but its luminous efficiency is limited, as has been shown in a previous chapter. When light is emitted by virtue of its temperature much invisible radiant energy accompanies the visible energy. The highest luminous efficiency attainable by pure temperature radiation will be reached when the temperature of a normal radiator reaches the vicinity of 10,000. to 11,000. The melting-points of metals are much lower than this. The tungsten filament in the most efficient lamps employing it is operating near its melting-point at the present time. Carbon is a most attractive element in respect to melting-point, for it melts at a temperature between 6000. and 7000. Even this is far below the most efficient temperature for the production of light by means of pure temperature radiation. There are possibilities of higher efficiency being obtained by operating arcs or filaments under pressure; however, it appears that highly efficient light of the future will result from a radical departure.

Scientists are becoming more and more intimate with the structure of matter. They are learning secrets every year which apparently are leading to a fundamental knowledge of the subject. When these mysteries are solved, who can say that man will not be able to create elements to suit his needs, or at least to alter the properties of the elements already available? If he could so alter the mechanism of radiation that a hot metal would radiate no invisible energy, he would have made a tremendous stride even in the production of light by virtue of high temperature. This property of selective radiation is possessed by some elements to a slight degree, but if treatment could enhance this property, luminous efficiency would be greatly increased. Certainly the principle of selectivity is a byway of possibilities.

A careful study of commonplace factors may result in a great step in light-production without the creation of new elements or compounds, just as such a procedure doubled the luminous efficiency of the tungsten filament when the gas-filled lamp appeared. There are a few elements still missing, according to the Periodic Law which has been so valuable in chemistry. When these turn up, they may be found to possess valuable properties for light-production; but this is a discouraging byway.

It is natural to inquire whether or not any mode of light-production now in use has a limiting luminous efficiency approaching the ultimate limit which is imposed by the visibility of radiation. The eye is able to convert radiant energy of different wave-lengths into certain fixed proportions of light. For example, radiant energy of such a wave-length as to excite the sensation of yellow-green is the most efficient and one watt of this energy is capable of being converted by the visual apparatus into about 625 lumens of light. Is this efficiency of conversion of the visual apparatus everlastingly fixed? For the answer it is necessary to turn to the physiologist, and doubtless he would suggest the curbing of the imagination. But is it unthinkable that the visual processes will always be beyond the control of man? However, to turn again to the physics of light-production, there are still several processes of producing light which are appealing.

Many years ago Geissler, Crookes, and other scientists studied the spectra of gases excited to incandescence by the electric discharge in so-called vacuum tubes. The gases are placed in transparent glass or quartz tubes at rather low pressures and a high voltage is impressed upon the ends of these tubes. When the pressure is sufficiently low, the gases will glow and emit light.

Twenty-eight years ago, D. McFarlan Moore developed the nitrogen tube, which was actually installed in various places. But there is such a loss of energy near the cathode that short "vacuum" tubes of this character are very inefficient producers of light. Efficiency is greatly increased by lengthening the tubes, so Moore used tubes of great length and a rather high voltage. As a tube of this sort is used, the gas gradually disappears and it must be replenished. In order to replenish the gas, Moore devised a valve for feeding gas automatically. An advantage of this mode of light-production is that the color or quality of the light may be varied by varying the gas used. Nitrogen yields a pinkish light; neon an orange light; and carbon dioxide a white light. Moore's carbon-dioxide tube is an excellent substitute for daylight and has been used for the discrimination of colors where this is an important factor. However, for this purpose devices utilizing color-screens which alter the light from the tungsten lamp to a daylight quality are being used extensively.

The vacuum-tube method of producing light has an advantage in the selection of a gas among a large number of possibilities, and some of the color effects of the future may be obtained by means of it. Claude has lately worked on light-production by vacuum tubes and has combined the neon tube with the mercury-vapor tube. The spectrum of neon to a large extent compensates for the absence of red light in the mercury spectrum, with a result that the mixture produces a more satisfactory light than that of either tube. However, this mode of light-production has not proved practicable in its present state of development. Fundamentally the limitations are those of the inherent spectral characteristics of gases. Doubtless the possibilities of the mechanisms of the tubes and of combining various gases have not been exhausted. Furthermore, if man ever becomes capable of controlling to some extent the structure of elements and of compounds, this method of light-production is perhaps more promising than others of the present day.

There is another attractive method of producing light and it has not escaped the writer of fiction. H. G. Wells, with his rare skill and with the license so often envied by the writer of facts, has drawn upon the characteristics of fluorescence and phosphorescence. In his story "The First Men in the Moon," the inhabitants of the moon illuminate their caverns by utilizing this phenomenon. A fluorescent liquid was prepared in large quantities. It emitted a brilliant phosphorescent glow and when it splashed on the feet of the earth-men it felt cold, but it glowed for a long time. This is a possibility of the future and many have had visions of such lighting. If the ceiling of a coal-mine was lined with glowing fireflies or with phosphorescent material excited in some manner, it would be possible to see fairly well with the dark-adapted eyes.

This leads to the class of phenomena included under the general term "luminescence." The definition of this term is not thoroughly agreed upon, but light produced in this manner does not depend upon temperature in the sense that a glowing tungsten filament emits light because it is sufficiently hot. A phosphorus match rubbed in the moist palm of the hand is seen to glow, although it is at an ordinary temperature. This may be termed "chemi-luminescence." Sidot blende, Balmain's paint, and many other compounds, when illuminated with ordinary light, and especially with ultra-violet and violet rays, will continue to glow for a long time. Despite their brightness they will be cold to the touch. This phenomenon would be termed "photo-luminescence," although it is better known as "phosphorescence." It should be noted that the latter term was carelessly originated, for phosphorus has nothing to do with it. The glow of the Geissler tube or electrically excited gas at low pressure would be an example of "electro-luminescence." The luminosity of various salts in the Bunsen-flame is due to so-called luminescence and there are many other examples of light-production which are included in the same general class. Inasmuch as light is emitted from comparatively cold bodies in these cases, it is popularly known as "cold" light.

There are many instances of light being emitted without being accompanied by appreciable amounts of invisible radiant energy and it is natural to hope for practical possibilities in this direction. As yet little is known regarding the efficiency of light-production by phosphorescence. The luminous efficiency of the radiant energy emitted by phosphorescent substances has been studied, but it seems strange that among the vast works on phosphorescent phenomena, scarcely any mention is made of the efficiency of producing light in this manner. For example, assume that phosphorescent zinc sulphide is excited by the light from a mercury-arc. All the energy falling upon it is not capable of exciting phosphorescence, as may be readily shown. Assuming that a known amount of radiant energy of a certain wave-length has been permitted to fall upon the phosphorescent material, then in the dark the latter may be seen to glow for a long time. An interesting point to investigate is the relation of the output to input; that is, the ratio of the total emitted light to the total exciting energy. This is a neglected aspect in the study of light-production by this means.

The firefly has been praised far and wide as the ideal light-source. It is an efficient radiator of light, for its light is "cold"; that is, it does not appear to be accompanied by invisible radiant energy. But little is said about its efficiency as a light-producer. Who knows how much fuel its lighting-plant consumes? The chemistry of light-production by living organisms is being unraveled and this part of the phenomenon will likely be laid bare before long. For an equal amount of energy radiated, the firefly emits a great many times more light than the most efficient lamp in use at the present time, but before the firefly is pronounced ideal, the efficiency of its light-producing process must be known.

There are many ways of exciting phosphorescence and fluorescence, the latter being merely an unenduring phosphorescence, which ceases when the exciting energy is cut off. Ultra-violet, violet, and blue rays are generally the most effective radiant energy for excitation purposes. X-rays and the high-frequency discharge are also powerful excitants. As already stated, virtually nothing is known of the efficiency of this mode of light-production or of the mechanism within the substance, but on the whole it is a remarkable phenomenon.

Radium is also a possibility in light-production and in fact has been practically employed for this purpose for several years. It or one of its compounds is mixed with a phosphorescent substance such as zinc sulphide and the latter glows continuously. Inasmuch as the life of some of the radium products is very long, such a method of illuminating watch-dials, scales of instruments, etc., is very practicable where they are to be read by eyes adapted to darkness and consequently highly sensitive to light. Whether or not radium will be manufactured by the ton in the future can only conjectured.

Owing to the limitations imposed by physical laws of radiation and by the physiological processes of vision the highest luminous efficiency obtainable by heating solid materials is only about 15 per cent. of the luminous efficiency of the most luminous radiant energy. At present there are no materials available which may be operated at the temperature necessary to reach even this efficiency. Great progress in the future of light-production as indicated by present knowledge appears to lie in the production of light which is unaccompanied by invisible radiant energy. At present such phenomena as fluorescence, phosphorescence, the light of the firefly, chemi-luminescence, etc., are examples of this kind of light-production. Of course, if science ever obtains control over the constitution of matter, many difficulties will disappear; for then man will not be dependent upon the elements and compounds now available but will be able to modify them to suit his needs.

XII

LIGHTING THE STREETS

In this age of brilliantly lighted boulevards and "great white ways" flooded with light from shop-windows, electric signs, and street-lamps, it is difficult to visualize the gloom which shrouded the streets a century ago. As the belated pedestrian walks along the suburban highways in comparative safety under adequate artificial lighting, he will realize the great influence of artificial light upon civilization if he recalls that not more than two centuries ago in London

it was a common practice ... that a hundred or more in a company, young and old, would make nightly invasions upon houses of the wealthy to the intent to rob them and that when night was come no man durst adventure to walk in the streets.

Inhabitants of the cities of the present time are inclined to think that crime is common on the streets at night, but what would it be without adequate artificial light? Two centuries ago in a city like London a smoking grease-lamp, a candle, or a basket of pine knots here and there afforded the only street-lighting, and these were extinguished by eleven o'clock. Lawlessness was hatched and hidden by darkness, and even the lantern or torch served more to mark the victim than to protect him. It has been said in describing the conditions of the age of dark streets that everybody signed his will and was prepared for death before he left his home. By comparison with the present, one is again encouraged to believe that the world grows better. Doubtless, artificial light projected into the crannies has had something to do with this change.

Adequate street-lighting is really a product of the twentieth century, but throughout the nineteenth century progress was steadily made from the beginning of gas-lighting in 1807. In preceding centuries crude lighting was employed here and there but not generally by the public authorities. In the earliest centuries of written history little is said of street-lighting. In those days man was not so much inclined to improve upon nature, beyond protecting himself from the elements, and he lighted the streets more as a festive outburst than as an economic proposition. Nevertheless, in the early writings occasionally there are indications that in the centers of advanced civilization some efforts were made to light the streets.

The old Syrian city of Antioch, which in the fourth century of the Christian era contained about four hundred thousand inhabitants, appears to have had lighted streets. Libanius, who lived in the early years of that century, wrote:

The light of the sun is succeeded by other lights, which are far superior to the lamps lighted by Egyptians on the festival of Minerva of Sais. The night with us differs from the day only in the appearance of the light; with regard to labor and employment, everything goes on well.

Although apparently labor was not on a strike, the soldiers caused disturbances, for in another passage he tells of riotous soldiers who

cut with their swords the ropes from which were suspended the lamps that afforded light in the night-time, to show that the ornaments of the city ought to give way to them.

Another writer in describing a dispute between two religious adherents of opposed creeds stated that they quarreled "till the streets were lighted" and the crowd of onlookers broke up, but not until they "spat in each other's face and retired." Thus it is seen that artificial light and civilization may advance, even though some human traits remain fundamentally unchanged.

Throughout the next thousand years there was little attempt to light the streets. Iron baskets of burning wood, primitive oil-lamps, and candles were used to some extent, but during all these centuries there was no attempt on the part of the government or of individuals to light the streets in an organized manner. In 1417 the Mayor of London ordained "lanthorns with lights to bee hanged out on the winter evenings betwixt Hallowtide and Candlemasse." This was during the festive season, so perhaps street-lighting was not the sole aim. Early in the sixteenth century, the streets of Paris being infested with robbers, the inhabitants were ordered to keep lights burning in the windows of all houses that fronted on the streets.

For about three centuries the citizens of London, and doubtless of Paris and of other cities, were reminded from time to time in official mandates "on pains and penalties to hang out their lanthorns at the appointed time." The watchman in long coat with halberd and lantern in hand supplemented these mandates as he made his rounds by,

A light here, maids, hang out your lights, And see your horns be clear and bright, That so your candle clear may shine, Continuing from six till nine; That honest men that walk along May see to pass safe without wrong.

In 1668, when some regulations were made for improving the streets of London, the inhabitants were ordered "for the safety and peace of the city to hang out candles duly to the accustomed hour." Apparently this method of obtaining lighting for the streets was not met by the enthusiastic support of the people, for during the next few decades the Lord Mayor was busy issuing threats and commands. In 1679 he proclaimed the "neglect of the inhabitants of this city in hanging and keeping out their lights at the accustomed hours, according to the good and ancient usage of this City and Acts of the Common Council on that behalf." The result of this neglect was "when nights darkened the streets then wandered forth the sons of Belial, flown with insolence and wine."

In 1694 Hemig patented a reflector which partially surrounded the open flame of a whale-oil lamp and possessed a hole in the top which aided ventilation. He obtained the exclusive rights of lighting London for a period of years and undertook to place a light before every tenth door, between the hours of six and twelve o'clock, from Michaelmas to Lady Day. His effort was a worthy one, but he was opposed by a certain faction, which was successful in obtaining a withdrawal of his license in 1716. Again the burden of lighting the streets was thrust upon the residents and fines were imposed for negligence in this respect. But this procedure after a few more years of desultory lighting was again found to be unsatisfactory.

In 1729 certain individuals contracted to light the streets of London by taxing the residents and paid the city for this monopoly. Householders were permitted to hang out a lantern or a candle or to pay the company for doing so. But robberies increased so rapidly that in 1736 the Lord Mayor and Common Council petitioned Parliament to erect lamps for lighting the city. An act was passed accordingly, giving them the privilege to erect lamps where they saw fit and to burn them from sunset to sunrise. A charge was made to the residents, on a sliding scale depending upon the rate of rental of the houses. As a consequence five thousand lamps were soon installed. In 1738 there were fifteen thousand street lamps in London and they were burned an average of five thousand hours annually.

In the annals of these early times street-lighting is almost invariably the result of an attempt to reduce the number of robberies and other crimes. In appealing for more street-lamps in 1744 the Lord Mayor and aldermen of London in a petition to the king, stated

that divers confederacies of great numbers of evil-disposed persons, armed with bludgeons, pistols, cutlasses, and other dangerous weapons, infest not only the private lanes and passages, but likewise the public streets and places of public concourse, and commit most daring outrages upon the persons of your Majesty's good subjects, whose affairs oblige them to pass through the streets, by terrifying, robbing and wounding them; and these facts are frequently perpetrated at such times as were heretofore deemed hours of security.

It has already been seen that gas-lighting was introduced in the streets of London for the first time in 1807. This marks the real beginning of public-service lighting companies. In the next decade interest in street-lighting by means of gas was awakened on the Continent, and it was not long before this new phase of civilization was well under way. Although this first gas-lighting was done by the use of open flames, it was a great improvement over all the preceding efforts. Lawlessness did not disappear entirely, of course, and perhaps it never will, but it skulked in the back streets. A controlling influence had now appeared.

But early innovations in lighting did not escape criticism and opposition. In fact, innovations to-day are not always received by unanimous consent. There were many in those early days who felt that what was good for them should be good enough for the younger generation. The descendants of these opponents are present to-day but fortunately in diminishing numbers. It has been shown that in Philadelphia in 1833 a proposal to install a gas-plant was met with a protest signed by many prominent citizens. A few paragraphs of an article entitled "Arguments against Light" which appeared in the Cologne Zeitung in 1816 indicate the character of the objections raised against street-lighting.

1 From the theological standpoint: Artificial illumination is an attempt to interfere with the divine plan of the world, which has preordained darkness during the night-time.

2 From the judicial standpoint: Those people who do not want light ought not to be compelled to pay for its use.

3 From the medical standpoint: The emanations of illuminating gas are injurious. Moreover, illuminated streets would induce people to remain later out of doors, leading to an increase in ailments caused by colds.

4 From the moral standpoint: The fear of darkness will vanish and drunkenness and depravity increase.

5 From the viewpoint of the police: The horses will get frightened and the thieves emboldened.

6 From the point of view of national economy: Great sums of money will be exported to foreign countries.

7 From the point of view of the common people: The constant illumination of streets by night will rob festive illuminations of their charm.

The foregoing objections require no comment, for they speak volumes pertaining to the thoughts and activities of men a century ago. It is difficult to believe that civilization has traveled so far in a single century, but from this early beginning of street-lighting social progress received a great impetus. Artificial light-sources were feeble at that time, but they made the streets safer and by means of them social intercourse was extended. The people increased their hours of activity and commerce, industry, and knowledge grew apace.

The open gas-jet and kerosene-flame lamps held forth on the streets until within the memory of middle-aged persons of to-day. The lamplighter with his ladder is still fresh in memory. Many of the towns and villages have never been lighted by gas, for they stepped from the oil-lamp to the electric lamp. The gas-mantle has made it possible for gas-lighting to continue as a competitor of electric-lighting for the streets.

In 1877 Mr. Brush illuminated the Public Square of Cleveland with a number of arc-lamps, and these met with such success that within a short time two hundred and fifty thousand open-arc lamps were installed in this country, involving an investment of millions of dollars. Adding to this investment a much greater one in central-station equipment, a very large investment is seen to have resulted from this single development in lighting.

This open-arc lamp was the first powerful light-source available and, appearing several years before the gas-mantle, it threatened to monopolize street-lighting. It consumed about 500 watts and had a maximum luminous intensity of about 1200 candles at an angle of about 45 degrees. Its chief disadvantage was its distribution of light, mainly at this angle of 45 degrees, which resulted in a spot of light near the lamp and little light at a distance. A satisfactory street-lighting unit must emit its light chiefly just below the horizontal in those cases where the lamps must be spaced far apart for economical reasons. On referring to the chapter on the electric arc it will be seen that the upper (positive) carbon of the open-arc emits most of the light. Thus most of the light tends to be sent downward, but the lower carbon obstructs some of this with a resulting dark spot beneath the lamp.

The gas-mantle followed closely after the arrival of the carbon arc and is responsible for the existence of gas-lighting on the streets at the present time. It is a large source of light and therefore its light cannot be controlled by modern accessories as well as the light from smaller sources, such as the arc or concentrated-filament lamp. As a consequence, there is marked unevenness of illumination along the streets unless the gas-mantle units are spaced rather closely. Even with the open-arc, without special light-controlling equipment there is about a thousand times the intensity near the lamps when placed on the corners of the block as there is midway between them.

In 1879 the incandescent filament lamp was introduced and it began to appear on the streets in a short time. It was a feeble, inefficient light-source, compared with the arc-lamp, but it had the advantage of being installed on a small bracket. As a consequence of simplicity of operation, the incandescent lamp was installed to a considerable extent, especially in the suburban districts.

The open-arc lamp possessed the disadvantage of emitting a very unsteady light and of consuming the carbons so rapidly that daily trimming was often necessary. In 1893 the enclosed arc appeared and although it consumed as much electrical energy as the open-arc and emitted considerably less light, it possessed the great advantage of operating a week without requiring a renewal of carbons. By surrounding the arc by means of a glass globe, little oxygen could come in contact with the carbons and they were not consumed very rapidly. The light was fairly steady and these arcs operated satisfactorily on alternating current. The latter feature simplified the generating and distributing equipment of the central station.

The magnetite or luminous arc-lamp next appeared and met with considerable success. It was more efficient than the preceding lamps but was handicapped by being solely a direct-current device. Those familiar with the generation and distribution of electricity will realize this disadvantage. However, its luminous intensity just below the horizontal was about 700 candles and its general distribution of light was fairly satisfactory. Later the flame-arcs began to appear and they were installed to some extent. The arc-lamp has served well in street-lighting from the year 1877, when the open-arc was introduced, until the present time, when the luminous-arc is the chief survivor of all the arc-lamps.

The carbon incandescent filament lamp was used extensively until 1909, when the tungsten filament lamp began to replace it very rapidly. However, it was not until 1914, when the gas-filled tungsten lamp appeared, that this type of light-source could compete with arc-lamps on the basis of efficiency. The helical construction of the filament made it possible to confine the filament of a high-intensity tungsten lamp in a small space and for the first time a high degree of control of the light of street lamps was possible. Prismatic "refractors" were designed, somewhat on the principle of the lighthouse refractor, so that the light would be emitted largely just below the horizontal. This type of distribution builds up the illumination at distant points between successive street lamps, which is very desirable in street-lighting. The incandescent filament lamp possesses many advantages over other systems. It is efficient; capable of subdivision; operates on direct and alternating current; requires little attention; and is capable of most successful use with light-controlling apparatus.

According to the reports of the Department of Commerce the number of electric arc-lamps for street-lighting supplied by public electric-light plants decreased from 348,643 in 1912 to 256,838 in 1917, while the number of electric incandescent filament lamps increased from 681,957 in 1912 to 1,389,382 in 1917.

Street-lighting is not only a reinforcement for the police but it decreases accidents and has come to be looked upon as an advertising medium. In the downtown districts the high-intensity "white-way" lighting is festive. The ornamental street lamps have possibilities in making the streets attractive and in illuminating the buildings. However, it is to be hoped that in the present age the streets of cities and towns will be cleared of the ragged equipment of the telephone and lighting companies. These may be placed in the alleys or underground, leaving the streets beautiful by day and glorified at night by the torches of advanced civilization.

XIII

LIGHTHOUSES

At the present time thousands of lighthouses, light-ships, and light-buoys guide the navigator along the waterways and into harbors and warn him of dangerous shoals. Many wonderful feats of engineering are involved in their construction and in no field of artificial lighting has more ingenuity been displayed in devising powerful beams of light. Many of these beacons of safety are automatic in operation and require little attention. It has been said that nothing indicates the liberality, prosperity, or intelligence of a nation more clearly than the facilities which it affords for the safe approach of the mariner to its shores. Surely these marine lights are important factors in modern navigation.

The first "lighthouses" were beacon-fires of burning wood maintained by priests for the benefit of the early commerce in the eastern part of the Mediterranean Sea. As early as the seventh century before Christ these beacon-fires were mentioned in writings. In the third century before the Christian era a tower said to be of a great height was built on a small island near Alexandria during the reign of Ptolemy II. The tower was named Pharos, which is the origin of the term "pharology" applied to the science of lighthouse construction. Caesar, who visited Alexandria two centuries later, described the Pharos as a "tower of great height, of wonderful construction." Fire was kept burning in it night and day and Pliny said of it, "During the night it appears as bright as a star, and during the day it is distinguished by the smoke." Apparently this tower served as a lighthouse for more than a thousand years. It was found in ruins in 1349. Throughout succeeding centuries many towers were built, but little attention was given to the development of light-sources and optical apparatus.

The first lighthouse in the United States and perhaps on the Western continents was the Boston Light, which was completed in 1716. A few days after it was put into operation a news item in a Boston paper heralded the noteworthy event as follows:

By virtue of an Act of Assembly made in the First Year of His Majesty's Reign, For Building and Maintaining a Light House upon the Great Brewster (called Beacon-Island) at the Entrance of the Harbour of Boston, in order to prevent the loss of the Lives and Estates of His Majesty's Subjects; the said Light House has been built; and on Fryday last the 14th Currant the Light was kindled, which will be very useful for all Vessels going out and coming in to the Harbour of Boston, or any other Harbours in the Massachusetts Bay, for which all Masters shall pay to the Receiver of Impost, one Penny per Ton Inwards, and another Penny Outwards, except Coasters, who are to pay Two Shillings each, at their clearance Out, And all Fishing Vessels, Wood Sloops, etc. Five Shillings each by the Year.

This was the practical result of a petition of Boston merchants made three years before. The tower was built of stone, at a cost of about ten thousand dollars. Two years later the keeper and his family were drowned and the catastrophe so affected Benjamin Franklin, a boy of thirteen, that he wrote a poem concerning it. The lighthouse was badly damaged during the Revolution, by raiding-parties, and in 1776, when the British fleet left the harbor, a squad of sailors blew it up. It was rebuilt in 1783 and has since been increased in height.

Apparently oil-lamps were used in it from the beginning, notwithstanding the fact that candles and coal fires served for years in many lighthouses of Europe. In 1789 sixteen lamps were used and in 1811 Argand lamps and reflectors were installed, with a revolving mechanism. It now ceased to be a fixed light and the day of flashing lights had arrived. At the present time the Boston Light emits a beam of 100,000 candle-power directed by modern lenses.

When the United States Government was organized in 1789 there were ten lighthouses owned by the Colonies, but the Boston Light was in operation thirty years before the others. Sandy Hook Light, New York Harbor, was established in 1764 and its original masonry tower is still standing and in use. It is the oldest surviving lighthouse in this country. It was built with funds raised by means of two lotteries authorized by the New York Assembly. A few days after it was lighted for the first time the following news item appeared in a New York paper:

On Monday evening last the New York Light-house erected at Sandy Hook was lighted for the first time. The House is of an Octagon Figure, having eight equal Sides; the Diameter at the Base 29 Feet; and at the top of the Wall, 15 Feet. The Lanthorn is 7 feet high; the Circumference 33 feet. The whole Construction of the Lanthorn is Iron; the Top covered with Copper. There are 48 Oil Blazes. The Building from the Surface is Nine Stories; the whole from Bottom to Top is 103 Feet.

From these early years the number of lighthouses has steadily grown, until now the United States maintains lights along 50,000 miles of coast-line and river channels, a distance equal to twice the circumference of the earth. It maintains at the present time about 15,000 aids to navigation at an annual cost of about $5,000,000. In 1916 this country was operating 1706 major lights, 53 light-ships, and 512 light-buoys--a total of 5323.

The earliest lighthouses were equipped with braziers or grates in which coal or wood was burned. These crude light-sources were used until after the advent of the nineteenth century and in one case until 1846. In the famous Eddystone tower off Plymouth, England, candles were used for the first time. The first Eddystone tower was completed in 1698, but it was swept away in 1703. Another was built and destroyed by fire in 1755. Smeaton then built another in 1759. Inasmuch as Smeaton is credited with having introduced the use of candles, this must have occurred in the eighteenth century; still it appears that, as we have said, the Boston Light, built in 1716, used oil-lamps from its beginning. However, Smeaton installed twenty-four candles of rather large size each credited with an intensity of 2.8 candles. The total luminous intensity of the light-source in this tower was about 67 candles. Inasmuch as this was before the use of efficient reflectors and lenses, it is obvious that the early lighthouses were rather feeble beacons.

According to British records, oil-lamps with flat wicks were first used in the Liverpool lighthouses in 1763. The Argand lamp, introduced in about 1784, became widely used. The better combustion obtained with this lamp having a cylindrical wick and a glass chimney greatly increased the luminous intensity and general satisfactoriness of the oil-lamp. Later Lange added an improvement by providing a contraction toward the upper part of the chimney. Rumford and also Fresnel devised multiple-wick burners, thus increasing the luminous intensity. In these early lamps sperm-oil and colza-oil were burned and they continued to be until the middle of the nineteenth century. Cocoanut-oil, lard-oil, and olive-oil have also been used in lighthouses.

Naturally, mineral oil was introduced as soon as it was available, owing to its lower cost; but it was not until nearly 1870 that a satisfactory mineral-oil lamp was in operation in lighthouses. Doty is credited with the invention of the first successful multiple-wick lighthouse lamp using mineral oil, and his lamp and modifications of it were very generally used until the latter part of the nineteenth century. These lamps are of two types--one in which oil is supplied to the burner under pressure and the other in which oil is maintained at a constant level. In some of the smallest lamps the ordinary capillarity of the wick is depended on to supply oil to the flame.

Coal-gas was introduced into lighthouses in about the middle of the nineteenth century. Inasmuch as the gas-mantle had not yet appeared, the gas was burned in jets. Various arrangements of the jets, such as concentric rings forming a stepped cone, were devised. The gas-mantle was a great boon to the mariner as well as to civilized beings in general. It greatly increases the intensity of light obtainable from a given amount of fuel and it is a fairly compact bright source which makes it possible to direct the light to some degree by means of optical systems. Owing to the elaborate apparatus necessary for making coal-gas, several other gases have been more desirable fuels for lighthouse lamps. Various simple gas-generators have been devised. Some of the high-flash mineral-oils are vaporized and burned under a mantle. Acetylene, which is so simply made by means of calcium carbide and water, has been a great factor in lighting for navigation. By the latter part of the nineteenth century lighthouses employing incandescent gas-burners were emitting beams of light having luminous intensities as great as several hundred thousand candles. These special gas-mantle light-sources have brightness as high as several hundred candles per square inch.

Electric arc-lamps were first introduced into lighthouse service in about 1860, but these lamps cannot be considered to have been really practicable until about 1875. In 1883 the British lighthouse authorities carried out an extensive investigation of arc-lamps. It was found that the whiter light from these lamps suffered a greater absorption by the atmosphere than the yellower light from oils, but the much greater luminous intensity of the arc-lamp more than compensated for this disadvantage. The final result of the investigation was the conclusion that for ordinary lighthouse purposes the oil-and gas-lamps were more suitable and economical than arc-lamps; but where great range was desired, the latter were much more advantageous, owing to their great luminous intensity. Electric incandescent filament lamps have been used for the less important lights, and recently there has been some application of the modern high-efficiency filament lamps.

Besides the high towers there are many minor beacons, light-ships, and light-buoys in use. Many of these are untended and therefore must operate automatically. The light-ship is used where it is impracticable or too expensive to build a lighthouse. Inasmuch as it is anchored in fairly deep water, it is safe in foggy weather to steer almost directly toward its position as indicated by the fog-signal. Light-ships are more expensive to maintain than lighthouses, but they have the advantages of smaller cost and of mobility; for sometimes it may be desired to move them. The first light-ship was established in 1732 near the mouth of the Thames, and the first in this country was anchored in Chesapeake Bay near Norfolk in 1820. The early ships had no mode of self-propulsion, but the modern ones are being provided with their own power. Oil and gas have been used as fuel for the light-sources and in 1892 the U. S. Lighthouse Board constructed a light-ship with a powerful electric light. Since that time several have been equipped with electric lights supplied by electric generators and batteries.

Untended lights were not developed until about 1880, when Pintsch introduced his welded buoys filled with compressed gas and thereby provided a complete lighting-plant. With improvements in lamps and controls the untended light-buoys became a success. The lights burn for several months, and even for a year continuously; and the oil-gas used appears to be very satisfactory. Recently some experiments have been made with devices which would be actuated by sunlight in such a manner that the light would be extinguished during the day excepting a small pilot-flame. By this means a longer period of burning without attention may be obtained. Electric filament lamps supplied by batteries or by cables from the shore have been used, but the oil-gas buoy still remains in favor. Acetylene has been employed as a fuel for light-buoys. Automatic generators have been devised, but the high-pressure system is more simple. In the latter case purified acetylene is held in solution under high pressure in a reservoir containing an asbestos composition saturated with acetone.

The light-sources of beacons have had the same history as those of other navigation lights. Many of these are automatic in operation, sometimes being controlled by clockwork. During the last twenty years the gas-mantle has been very generally applied to beacon-lights. In the latter part of the nineteenth century a mineral-oil lamp was devised with a permanent wick made by forming upon a thick wick a coating of carbon. The operation is such that this is not consumed and it prevents further burning of the wick.

The optical apparatus of navigation lights has undergone many improvements in the past century. The early lights were not equipped with either reflecting or refracting apparatus. In 1824 Drummond devised a scheme for reflecting light in order that a distant observer might make a reading upon the point where the apparatus was being operated by another person. He was led by his experiments to suggest the application of mirrors to lighthouses. His device was essentially a parabolic mirror similar to the reflectors now widely used in automobile head-lamps, search-lights, etc. He employed the lime-light as a source of light and was enthusiastic over the results obtained. His discussion published in 1826 indicates that little practical work had been done up to that time toward obtaining beams or belts of light by means of optical apparatus. However, lighthouse records show that as early as 1763 small silvered plane glasses were set in plaster of Paris in such a manner as to form a partially enveloping reflector. Spherical reflectors were introduced in about 1780 and parabolic reflectors about ten years later.

All the earlier lights were "fixed," but as it is desirable that the mariner be able to distinguish one light from another, the revolving mechanism evolved. By its agency characteristic flashes are obtained and from the time interval the light is recognized. The first revolving mechanism was installed in 1783. The early flashing lights were obtained by means of revolving reflectors which gathered the light and directed it in the form of a beam or pencil. The type of parabolic reflector now in use does not differ essentially from that of an automobile head-lamp, excepting that it is larger.

Lenses appear to have been introduced in the latter part of the nineteenth century. They were at first ground from a solid piece of glass, in concentric zones, in order to reduce the thickness. They were similar in principle to some of the tail-light lenses used at present on automobiles. Later the lenses were built up by means of separate annular rings. The name of Fresnel is permanently associated with lighthouse lenses because in 1822 he developed an elaborate built-up lens of annular rings. The centers of curvature of the different rings receded from the axis as their distance from the center increased, in such a manner as to overcome a serious optical defect known as spherical aberration. Fresnel devised many improvements in which he used refracting and reflecting prisms for the outer elements.

The optical apparatus of lighthouses usually aims (1) to concentrate the rays of light into a pencil of light, (2) to concentrate them into a belt of light, or (3) to concentrate the rays over a limited azimuth. In the first case a single lens or a parabolic reflector suffices, but in the second case a cylindrical lens which condenses the light vertically into a horizontal sheet of light is essential. The third case is a combination of the first two. The modern lighthouse lenses are very elaborate in construction, being built up by means of many elements into several sections. For example, the central section may consist of a spherical lens ground with annular rings. In the next section refracting prisms may be used and in the outer section reflecting glass prisms are employed. The various elements are carefully designed according to the laws of geometrical optics.

The flashing light has such advantages over the fixed that it is generally used for important beacons. A variety of methods of obtaining intermittent light have been employed, but they are not of particular interest. Sometimes the lens or reflector is revolved and in other types an opaque screen containing slits is revolved. In the larger lighthouses the optical apparatus and its structure sometimes weigh several tons. When it is necessary to revolve apparatus of this weight, the whole mechanism is floated upon mercury contained in a cast-iron vessel of suitable size, and by an ingenious arrangement only a small portion of mercury is required.

The characteristics of navigation lights are varied considerably in order to enable the mariner to distinguish them and thereby to learn exactly where he is. The fixed light is liable to be confused with others, so it has now become a minor light. Flashes of short duration followed by longer periods of darkness are extensively used. The mariner by timing the intervals is able to recognize the light. This method is extended to groups of short flashes followed by longer intervals of darkness. In fact, short flashes have been employed to indicate a certain number so that a mariner could recognize the light by a number rather than by means of his watch. However, a time element is generally used. A combination of fixed light upon which is superposed a flash or a group of flashes of white or of colored light has been used, but it is in disrepute as being unreliable. A type known as "occulating lights" consists of a fixed light which is momentarily eclipsed, but the duration of the eclipse is usually less than that of the light. Obviously, groups of eclipses may be used. Sometimes lights of different colors are alternated without any dark intervals. The colored ones used are generally red and green, but these are short-range lights at best. Colored sectors are sometimes used over portions of the field, in order to indicate dangers, and white light shows in the fairway. These are usually fixed lights for marking the channel.

The distance at which a light may be seen at sea depends upon its luminous intensity, upon its color or spectral composition, upon its height and that of the observer's eyes above the sea-level, and upon the atmospheric conditions. Assuming a perfectly clear atmosphere, the visibility of a light-source apparently depends directly upon its candle-power. The atmosphere ordinarily absorbs the red, orange, and yellow rays less than the green, blue, and violet rays. This is demonstrated by the setting sun, which as it approaches closer to the horizon changes from yellow to orange and finally to red as the amount of atmosphere between it and the eye increases. For this reason a red light would have a greater range than a blue light of the same luminous intensity.

Under ordinary atmospheric conditions the range of the more powerful light-sources used in lighthouses is greater than the range as limited by the curvature of the earth. For the uncolored illuminants the range in nautical miles appears to be at least equal to the square root of the candle-power. A real practical limitation which still exists is the curvature of the earth, and the distance an object may be seen by the eye at sea-level depends upon the height of the object. The relation is approximately expressed thus,--

Range in nautical miles = 8/7 square root of Height of object in feet. For example, the top of a tower 100 feet high is visible to an eye at sea-level a distance of 8/7 square root of 100 = 80/7 = 11.43 miles. Now if the eye is 49 feet above sea-level, a similar computation will show how far away it may be seen by the original eye at sea-level. This is 8/7 square root of 49 = 8 miles. Hence an eye 49 feet above sea-level will be able to see the top of the 100-foot tower at a distance of 11.43 + 8 or 19.43 nautical miles. Under these conditions an imaginary line drawn from the top of the tower to the eye will be just tangent to the spherical surface of the sea at a distance of 8 miles from the eye and 11.43 miles from the tower.

The luminous intensity of a light-source or of the beam of light is directly responsible for the range. The luminous intensity of the early beacon-fires and oil-lamps was equivalent to a few candles. The improvements in light-sources and also in reflecting and refracting optical systems have steadily increased the candle-power of the beams, until to-day the beams from gas-lamps have intensities as high as several hundred thousand candle-power.

The beams sent forth by modern lighthouses equipped with electric lamps and enormous light-gathering devices are rated in millions of candle-power. In fact, Navesink Light at the entrance of New York Bay is rated as high as 60,000,000 candle-power.

Of course, light-production has increased enormously in efficiency in the past century, but without optical devices for gathering the light, the enormous beam intensity would not be obtained. For example, consider a small source of light possessing a luminous intensity of one candle in all directions. If all this light which is emitted in all directions is gathered and sent forth in a beam of small angle, say one thousandth of the total angle surrounding a point, the intensity of this beam would be 1000 candles. It is in this manner that the enormous beam intensities are built up.

There is an interesting point pertaining to short flashes of light. To the dark-adapted eye a brief flash is registered as of considerably higher intensity than if the light remained constant. In other words, the lookout on a vessel is adapted to darkness and a flash from a beam of light is much brighter than if the same beam were shining steadily. This is a physiological phenomenon which operates in favor of the flashing light.

Doubtless, the reader has noted that reliability, simplicity, and low cost of operation are the primary considerations for light-sources used as aids to navigation. This accounts for the continued use of oil and gas. From an optical standpoint the electric arc-lamps and concentrated-filament lamps are usually superior to the earlier sources of light, but the complexity of a plant for generating electricity is usually a disadvantage in isolated places. The larger light-ships are now using electricity generated by apparatus installed in the vessels. There seems to be a tendency toward the use of more buoys and fewer lighthouses, but the beam-intensities of the latter are increasing.

In the hundred years since the Boston Light was built the same great changes wrought by the development of artificial light in other activities of civilization have appeared in the beacons of the mariner. The development of these aids to navigation has been wonderful, but it must go on and on. The surface of the earth comprises 51,886,000 square statute miles of land and 145,054,000 square miles of water. Three fourths of the earth's surface is water and the oceans will always be highways of world commerce. All the dangers cannot be overcome, but human ingenuity is capable of great achievements. Wreckage will appear along the shore-lines despite the lights, but the harvest of the shoals has been much reduced since the time described by Robert Louis Stevenson, when the coast people in the Orkneys looked upon wrecks as a source of gain. He states:

It had become proverbial with some of the inhabitants to observe that "if wrecks were to happen, they might as well be sent to the poor island of Sanday as anywhere else." On this and the neighboring island, the inhabitants have certainly had their share of wrecked goods. On complaining to one of the pilots of the badness of his boat's sails, he replied with some degree of pleasantry, "Had it been His [God's] will that you come na here wi these lights, we might a' had better sails to our boats and more o' other things."

In the leasing of farms, a location with a greater probability of shipwreck on the shore brought a much higher rent.

XIV

ARTIFICIAL LIGHT IN WARFARE

When the recent war broke out science responded to the call and under the stress of feverish necessity compressed the normal development of a half-century into a few years. The airplane, in 1914 a doubtful plaything of daredevils, emerged from the war a perfected thing of the air. Lighting did not have the glamor of flying or the novelty of chemical warfare, but it progressed greatly in certain directions and served well. While artificial lighting conducted its unheralded offensive by increasing production in the supporting industries and helped to maintain liaison with the front-line trenches by lending eyes to transportation, it was also doing its part at the battle front. Huge search-lights revealed the submarine and the aerial bomber; flares exposed the manoeuvers of the enemy; rockets brought aid to beleaguered vessels and troops; pistol lights fired by the aerial observer directed artillery fire; and many other devices of artificial light were in the fray. Many improvements were made in search-lights and in signaling devices and the elements of the festive fireworks of past ages were improved and developed for the needs of modern warfare.

Night after night along the battle front flares were sent up to reveal patrols and any other enemy activity. On the slightest suspicion great swarms of these brilliant lights would burst forth as though flocks of huge fireflies had been disturbed. They were even used as light barrages, for movements could be executed in comparative safety when a large number of these lights lay before the enemy's trenches sputtering their brilliant light. The airman dropped flares to illuminate his target or his landing field. The torches of past parades aided the soldier in his night operations and rockets sent skyward radiated their messages to headquarters in the rear. The star-shell had the same missions as other flares, but it was projected by a charge of powder from a gun. These and many modifications represent the useful applications of what formerly were mere "fireworks." Those which are primarily signaling devices are discussed in another chapter, but the others will be described sufficiently to indicate the place which artificial light played in certain phases of warfare.

The illuminating compounds used in these devices are not particularly new, consisting essentially of a combustible powder and chemical salts which make the flame luminous and give it color when desired. Among the ingredients are barium nitrate, potassium perchlorate, powdered aluminum, powdered magnesium, potassium nitrate, and sulphur. One of the simplest mixtures used by the English is,

Barium nitrate 37 per cent. Powdered magnesium 34 per cent. Potassium nitrate 29 per cent.

The magnesium is coated with hot wax or paraffin, which not only acts as a binder for the mixture when it is pressed into its container but also serves to prevent oxidation of the magnesium when the shells are stored. The barium and potassium nitrates supply the oxygen to the magnesium, which burns with a brilliant white flame. The potassium nitrate takes fire more readily than the barium nitrate, but it is more expensive than the latter.

Owing to the cost of magnesium, powdered aluminum has been used to some extent as a substitute. Aluminum does not have the illuminating value of magnesium and it is more difficult to ignite, but it is a good substitute in case of necessity. An English mixture containing these elements is,

Barium nitrate 58 per cent. Magnesium 29 per cent. Aluminum 13 per cent.

Mixtures which are slow to ignite must be supplemented by a primary mixture which is readily ignited. For obtaining colored lights it is only necessary to add chemicals which will give the desired color. The mixtures can be proportioned by means of purely theoretical considerations; that is, just enough oxygen can be present to burn the fuel completely. However, usually more oxygen is supplied than called for by theory.

The illuminating shell is perhaps the most useful of these devices to the soldier. It has been constructed with and without parachutes, the former providing an intense light for a brief period because it falls rapidly. These shells of the larger calibers are equipped with time-fuses and are generally rather elaborate in construction. The shell is of steel, and has a time-fuse at the tip. This fuse ignites a charge of black powder in the nose of the shell and this explosion ejects the star-shell out of the rear of the steel casing. At the same time the black powder ignites the priming mixture next to it, which in turn ignites the slow-burning illuminating compound. The star-shell has a large parachute of strong material folded in the rear of the casing and the cardboard tube containing the illuminating mixture is attached to it. The time of burning varies, but is ordinarily less than a minute. Certain structural details must be such as to endure the stresses of a high muzzle velocity. Furthermore, a velocity of perhaps 1000 feet per second still obtains when the star-shell with its parachute is ejected at the desired point in the air.

The non-parachute illuminating shell is designed to give an intense light for a brief interval and is especially applicable to defense against air raids. Such a light aims to reveal the aircraft in order that the gunners may fire at it effectively. These shells are fitted with time-fuses which fire the charge of black powder at the desired interval after the discharge of the shell from the gun. The contents of the shell are thereby ejected and ignited. The container for the illuminating material is so designed that there is rapid combustion and consequently a brilliant light for about ten seconds. The enemy airman in this short time is unable to obtain any valuable knowledge pertaining to the earth below and furthermore he is likely to be temporarily blinded by the brilliant light if it is near him.

The rifle-light which resembles an ordinary rocket, is fired from a rifle and is designed for short-range use. It consists of a steel cylindrical shell a few inches long fastened to a steel rod. A parachute is attached to the cardboard container in which the illuminating mixture is packed and the whole is stowed away in the steel shell. Shore delay-fuses are used for starting the usual cycle of events after the rifle-light has been fired from the gun. The steel rod is injected into the barrel of a rifle and a blank cartridge is used for ejecting this rocket-like apparatus. Owing to inertia the firing-pin in the shell operates and the short delay-fuse is thus fired automatically an instant after the trigger of the rifle is pulled.

Illuminating "bombs" of the same general principles are used by airmen in search of a landing for himself or for a destructive bomb; in signaling to a gunner, and in many other ways. They are simple in construction because they need not withstand the stresses of being fired from a gun; they are merely dropped from the aircraft. The mechanism of ignition and the cycle of events which follow are similar to those of other illuminating shells.

The value of such artificial-lighting devices depends both upon luminous intensity and time of burning. Although long-burning is not generally required in warfare, it is obvious that more than a momentary light is usually needed. In general, high candle-power and long-burning are opposed to each other, so that the most intense lights of this character usually are of short duration. Typical performances of two flares of the same composition are as follows:

	Flare No. 1	Flare No. 2
Average candle-power	270,000	95,000
Seconds of burning	10	35
Candle-seconds	2,700,000	3,325,000
Cubic inches of compound	6	7
Candle-seconds per cubic inch	450,000	475,000
Candle-hours per cubic inch	125	132

The illuminating compound was the same in these two flares, which differed only in the time allowed for burning. Of course, the measurements of the luminous intensity of such flares is difficult because of the fluctuations, but within the errors of the measurements it is seen that the illuminating power of the compound is about the same regardless of the time of burning. The light-source in the case of burning powders is really a flame, and inasmuch as the burning end hangs downward, more light is emitted in the lower hemisphere than in the upper. The candle-power of the largest flares equals the combined luminous intensities of 200 street arc-lamps or of 10,000 ordinary 40-watt tungsten lamps such as are used in residence lighting.

It is interesting to note the candle-hours obtained per cubic inch of compound and to find that the cost of this light is less than that of candles at the present time and only five or ten times greater than that of modern electric lighting.

Illuminating shells in use during the recent war were designed for muzzle velocities as high as 2700 feet per second and were gaged to ignite at any distance from a quarter of a mile to several miles. The maximum range of illuminating shells fired from rifles was about 200 yards; for trench mortars about one mile; and from field and naval guns about four miles.

The search-light has long been a valuable aid in warfare and during the recent conflict considerable attention was given to its development and application. It is used chiefly for detecting and illuminating distant targets, but this covers a wide range of conditions and requirements. In order that a search-light may be effective at a great distance, as much as possible of the light emitted by a source is directed into a beam of light of as nearly parallel rays as can be obtained. Reflectors are usually employed in military search-lights, and in order that the beam may be as nearly parallel (minimum divergence) as possible, the light must be emitted by the smallest source compatible with high intensity. This source is placed at the proper point in respect to a large parabolic reflecter which renders the rays parallel or nearly so.

Ever since its advent the electric arc has been employed in large search-lights, with which the army and the navy were supplied; however, the greatest improvements have been made under the stress of war. The science of aeronautics advanced so rapidly during the recent war that the necessity for powerful search-lights was greatly augmented and as the conflict progressed the enemy airmen came to look upon the newly developed ones with considerable concern. The rapidly moving aircraft and its high altitude brought new factors into the design of these lights. It now became necessary to have the most intense beam and to be able to sweep the heavens with it by means of delicate controlling apparatus, for the targets were sometimes minute specks moving at high speed at altitudes as high as five miles. Furthermore, owing to the shifting battle areas, mobile apparatus was necessary.

The control of light by means of reflectors has been studied for centuries, but until the advent of the electric arc the light-sources were of such large areas that effective control was impossible. Optical devices generally are considered in connection with "point sources," but inasmuch as no light can be obtained from a point, a source of small dimensions and of high brightness is the most effective compromise. Parabolic mirrors were in use in the eighteenth century and their properties were known long before the first search-light worthy of the name was made in 1825 by Drummond, who used as a source of light a piece of lime heated to incandescence in a blast flame. He finally developed the "lime-light" by directing an oxyhydrogen flame upon a piece of lime and this device was adapted to search-lights and to indoor projection. It is said that the first search-light to be used in warfare was a Drummond lime-light which played a part in the attack on Fort Wagner at Charleston in 1863.

In 1848 the first electric arc lamp used for general lighting was installed in Paris. It was supplied with current by a large voltaic cell, but the success of the electric arc was obliged to await the development of a more satisfactory source of electricity. A score of years was destined to elapse, after the public was amazed by the first demonstration, before a suitable electric dynamo was invented. With the advent of the dynamo, the electric arc was rapidly developed and thus there became available a concentrated light-source of high intensity and great brilliancy. Gradually the size was increased, until at the present time mirrors as large as seven feet in diameter and electric currents as great as several hundred amperes are employed. The beam intensities of the most powerful search-lights are now as great as several hundred million candles.

The most notable advance in the design of arc search-lights was achieved in recent years by Beck, who developed an intensive flame carbon-arc. His chief object was to send a much greater current through the arc than had been done previously without increasing the size of the carbons and the unsteadiness of the arc. In the ordinary arc excessive current causes the carbons to disintegrate rapidly unless they are of large diameter. Beck directed a stream of alcohol vapor at the arc and they were kept from oxidizing. He thus achieved a high current-density and much greater beam intensities. He also used cored carbons containing certain metallic salts which added to the luminous intensity, and by rotation of the positive carbon so that the crater was kept in a constant position, greater steadiness and uniformity were obtained. Tests show that, in addition to its higher luminous efficiency, an arc of this character directs a greater percentage of the light into the effective angle of the mirror. The small source results in a beam of small divergence; in other words, the beam differs from a cylinder by only one or two degrees. If the beam consisted entirely of parallel rays and if there were no loss of light in the atmosphere by scattering or by absorption, the beam intensity would be the same throughout its entire length. However, both divergence and atmospheric losses tend to reduce the intensity of the beam as the distance from the search-light increases.

Inasmuch as the intensity of the beam depends upon the actual brightness of the light-source, the brightness of a few modern light-sources are of interest. These are expressed in candles per square inch of projected area; that is, if a small hole in a sheet of metal is placed next to the light-source and the intensity of the light passing through this hole is measured, the brightness of the hole is easily determined in candles per square inch.

BRIGHTNESS OF LIGHT-SOURCES IN CANDLES PER SQUARE INCH

Kerosene flame 5 to 10 Acetylene 30 to 60 Gas-mantle 30 to 500 Tungsten filament (vacuum) lamp 750 to 1,200 Tungsten filament (gas-filled) lamp 3,500 to 18,000 Magnetite arc 4,000 to 6,000 Carbon arc for search-lights 80,000 to 90,000 Flame arc for search-lights 250,000 to 350,000 Sun (computed mean) about 1,000,000

As the reflector of a search-light is an exceedingly important factor in obtaining high beam-intensities, considerable attention has been given to it since the practicable electric arc appeared. The parabolic mirror has the property of rendering parallel, or nearly so, the rays from a light-source placed at its focus. If the mirror subtends a large angle at the light-source, a greater amount of light is intercepted and rendered parallel than in the case of smaller subtended angles; hence, mirrors are large and of as short focus as practicable. Search-light projectors direct from 30 to 60 per cent. of the available light into the beam, but with lens systems the effective angle is so small that a much smaller percentage is delivered in the beam. Mangin in

1874 made a reflector of glass in which both outer and inner surfaces were spherical but of different radii of curvature, so that the reflector was thicker in the middle. This device was "silvered" on the outside and the refraction in the glass, as the light passed through it to the mirror and back again, corrected the spherical aberration of the mirrored surface. These have been extensively used. Many combinations of curved surfaces have been developed for special projection purposes, but the parabolic mirror is still in favor for powerful search-lights. The tip of the positive carbon is placed at its focus and the effective angle in which light is intercepted by the mirror is generally about 125 degrees. Within this angle is included a large portion of the light emitted by the light-source in the case of direct-current arcs. If this angle is increased for a mirror of a given diameter by decreasing its focal length, the divergence of the beam is increased and the beam-intensity is diminished. This is due to the fact that the light-source now becomes apparently larger; that is, being of a given size it now subtends a larger angle at the reflector and departs more from the theoretical point.

When the recent war began the search-lights available were intended generally for fixed installations. These were "barrel" lights with reflectors several feet in diameter, the whole output sometimes weighing as much as several tons. Shortly after the entrance of this country into the war, a mobile "barrel" search-light five feet in diameter was produced, which, complete with carriage, weighed only 1800 pounds. Later there were further improvements. An example of the impetus which the stress of war gives to technical accomplishments is found in the development of a particular mobile searchlight. Two months after the War Department submitted the problems of design to certain large industrial establishments a new 60-inch search-light was placed in production. It weighed one fifth as much as the previous standard; it had one twentieth the bulk; it was much simpler; it could be built in one fourth the time; and it cost half as much. Remote control of the apparatus has been highly developed in order that the operator may be at a distance from the scattered light near the unit. If he is near the search-light, this veil of diffused light very seriously interferes with his vision.

Mobile power-units were necessary and the types developed used the automobile engine as the prime mover. In one the generator is located in front of the engine and supported beyond the automobile chassis. In another type the generator is located between the automobile transmission and the differential. A standard clutch and gear-shift lever is employed to connect the engine either with the generator or with the propeller shaft of the truck. The first type included a 115-volt, 15-kilowatt generator, a 36-inch wheel barrel search-light, and 500 feet of wire cable. The second type included a 105-volt, 20-kilowatt generator, a 60-inch open searchlight, and 600 feet of cable. This type has been extended in magnitude to include a 50-kilowatt generator. When these units are moved, the search-light and its carriage are loaded upon the rear of the mobile generating equipment. An idea of the intensities obtainable with the largest apparatus is gained from illumination produced at a given distance. For example, the 15-kilowatt search-light with highly concentrated beam, produced an illumination at 930 feet of 280 foot-candles. At this point this is the equivalent of the illumination produced by a source having a luminous intensity of nearly 250,000,000 candles.

Of course, the range at which search-lights are effective is the factor of most importance, but this depends upon a number of conditions such as the illumination produced by the beam at various distances, the atmospheric conditions, the position of the observer, the size, pattern, color, and reflection-factor of the object, and the color, pattern, and reflection-factor of the background. These are too involved to be discussed here, but it may be stated that under ordinary conditions these powerful lights are effective at distances of several miles. According to recent work, it appears that the range of a search-light in revealing a given object under fixed conditions varies about as the fourth root of its intensity.

Although the metallic parabolic reflector is used in the most powerful search-lights, there have been many other developments adapted to warfare. Fresnel lenses have been used above the arc for search-lights whose beams are directed upward in search of aircraft, thus replacing the mirror below the arc, which, owing to its position, is always in danger of deterioration by the hot carbon particles dropping upon it. For short ranges incandescent filament lamps have been used with success. Oxyacetylene equipment has found application, owing to its portability. The oxyacetylene flame is concentrated upon a small pellet of ceria, which provides a brilliant source of small dimensions. A tank containing about 1000 liters of dissolved acetylene and another containing about 1100 liters of oxygen supply the fuel. A beam having an intensity of about 1,500,000 candles is obtained with a consumption of 40 liters of each of the gases per hour. At this rate the search-light may be operated twenty hours without replenishing.

Although the beacon-light for nocturnal airmen is a development which will assume much importance in peaceful activities, it was developed chiefly to meet the requirements of warfare. These do not differ materially from those which guide the mariner, except that the traveler in the ocean is far above the plane on which the beacon rests. For this reason the lenses are designed to send light generally upward. In foreign countries several types of beacons for navigation have been in use. In one the light from the source is freely emitted in all upward directions, but the light normally emitted into the lower hemisphere is turned upward by means of prisms. In a more elaborate type, belts of lenses are arranged so as to send light in all directions above the horizontal plane. A flashing apparatus is used to designate the locality by the number or character of the flashes. Electric filaments and acetylene flames have been used as the light-sources for this purpose. In another type the light is concentrated in one azimuth and the whole beacon is revolved. Portable beacons employing gas were used during the war on some of the flying-fields near the battle front.

All kinds of lighting and lighting-devices were used depending upon the needs and material available. Even self-luminous paint was used for various purposes at the front, as well as for illuminating watch-dials and the scales of instruments. Wooden buttons two or three inches in diameter covered with self-luminous paint could be fixed wherever desired and thus serve as landmarks. They are visible only at short distances and the feebleness of their light made them particularly valuable for various purposes at the battle front. They could be used in the hand for giving optical signals at a short distance where silence was essential. Self-luminous arrows and signs directed troops and trucks at night and even stretcher-bearers have borne self-luminous marks on their backs in order to identify them to their friends.

Somewhat analogous to this application of luminous paint is the use of blue light at night on battle-ships and other vessels in action or near the enemy. Several years ago a Brazilian battle-ship built in this country was equipped with a dual lighting-system. The extra one used deep-blue light, which is very effective for eyes adapted to darkness or to very low intensities of illumination and is a short-range light. Owing to the low luminous intensity of the blue lights they do not carry far; and furthermore, it is well established that blue light does not penetrate as far through ordinary atmosphere as lights of other colors of the same intensity.

The war has been responsible for great strides in certain directions in the development and use of artificial light and the era of peace will inherit these developments and will adapt them to more constructive purposes.

XV

SIGNALING

From earliest times the beacon-fire has sent forth messages from hilltops or across inaccessible places. In this country, when the Indian was monarch of the vast areas of forest and prairie, he spread news broadcast to roving tribesmen by means of the signal-fire, and he flashed his code by covering and uncovering it. Castaways, whether in fiction or in reality, instinctively turn to the beacon-fire as a mode of attracting a passing ship. On every hand throughout the ages this simple means of communication has been employed; therefore, it is not surprising that mankind has applied his ingenuity to the perfection of signaling by means of light, which has its own peculiar fields and advantages. Of course, wireless telephony and telegraphy will replace light-signaling to some extent, but there are many fields in which the last-named is still supreme. In fact, during the recent war much use was made of light in this manner and devices were developed despite the many other available means of signaling. One of the chief advantages of light as a signal is that it is so easily controlled and directed in a straight line. Wireless waves, for example, are radiated broadcast to be intercepted by the enemy.

The beginning of light-signaling is hidden in the obscurity of the past. Of course, the most primitive light-signals were wood fires, but it is likely that man early utilized the mirror to reflect the sun's image and thus laid the foundation of the modern heliograph. The Book of Job, which is probably one of the oldest writings available, mentions molten mirrors. The Egyptians in the time of Moses used mirrors of polished brass. Euclid in the third century before the Christian era is said to have written a treatise in which he discussed the reflection of light by concave mirrors. John Peckham, Archbishop of Canterbury in the thirteenth century, described mirrors of polished steel and of glass backed with lead. Mirrors of glass coated with an alloy of tin and mercury were made by the Venetians in the sixteenth century. Huygens in the seventeenth century studied the laws of refraction and reflection and devised optical apparatus for various purposes. However, it was not until the eighteenth century that any noteworthy attempts were made to control artificial light for practical purposes. Dollond in 1757 was the first to make achromatic lenses by using combinations of different glasses. Lavoisier in 1774 made a lens about four feet in diameter by constructing a cell of two concave glasses and filling it with water and other liquids. It is said that he ignited wood and melted metals by concentrating the sun's image upon them by means of this lens. About that time Buffon made a built-up parabolic mirror by means of several hundred small plane mirrors set at the proper angles. With this he set fire to wood at a distance of more than two hundred feet by concentrating the sun's rays. He is said also to have made a lens from a solid piece of glass by grinding it in concentric steps similar to the designs worked out by Fresnel seventy years later. These are examples of the early work which laid the foundation for the highly perfected control of light of the present time.

While engaged in the survey of Ireland, Thomas Drummond in 1826 devised apparatus for signaling many miles, thus facilitating triangulation. Distances as great as eighty miles were encountered and it appeared desirable to have some method for seeing a point at these great distances. Gauss in 1822 used the reflection of the sun's image from a plane mirror and Drummond also tried this means. The latter was successful in signaling 45 miles to a station which because of haze could not be seen, or even the hill upon which it rested. Having demonstrated the feasibility of the plan, he set about making a device which would include a powerful artificial light in order to be independent of the sun. In earlier geodetic surveys Argand lamps had been employed with parabolic reflectors and with convex lenses, but apparently these did not have a sufficient range. Fresnel and Arago constructed a lens consisting of a series of concentric rings which were cemented together, and on placing this before an Argand lamp possessing four concentric wicks, they obtained a light which was observed at forty-eight miles.

Despite these successes, Drummond believed the parabolic mirror and a more powerful light-source afforded the best combination for a signal-light. In searching for a brilliant light-source he experimented with phosphorus burning in oxygen and with various brilliant pyrotechnical preparations.

However, flames were unsteady and generally unsuitable. He then turned in the direction which led to his development of the lime-light. In his first apparatus he used a small sphere of lime in an alcohol flame and directed a jet of oxygen through the flame upon the lime. He thereby obtained, according to his own description in 1826,

a light so intense that when placed in the focus of a reflector the eye could with difficulty support its splendor, even at a distance of forty feet, the contour being lost in the brilliancy of the radiation.

He then continued to experiment with various oxides, including zirconia, magnesia, and lime from chalk and marble. This was the advent of the lime-light, which should bear Drummond's name because it was one of the greatest steps in the evolution of artificial light.

By means of this apparatus in the survey, signals were rendered visible at distances as great as one hundred miles. Drummond proposed the use of this light-source in the important lighthouses at that time and foresaw many other applications. The lime-light eventually was extensively used as a light-signaling device. The heliograph, which utilizes the sun as a light-source, has been widely used as a light-signaling apparatus and Drummond perhaps was the first to utilize artificial light with it. The disadvantage of the heliograph is the undependability of the sun. With the adoption of artificial light, various optical devices have come into use.

Philip Colomb perhaps is deserving of the credit of initiating modern signaling by flashing a code. He began work on such a system in 1858 and as an officer in the British Navy worked hard to introduce it. Finally, in 1867, the British Navy adopted the flashing-system, in which a light-source is exposed and eclipsed in such a manner as to represent dots and dashes analogous to the Morse code. At first the rate of transmission of words was from seven to ten per minute. Recently much more sensitive apparatus is available, and with such devices the rate is limited only by the sluggishness of the visual process. This initial system was very successful in the British Navy and it was soon found that a fleet could be handled with ease and safety in darkness or in fog. Inasmuch as the "dot-and-dash" system requires only two elements, it may be transmitted by various means. A lantern may be swung in short and long arcs or dipped accordingly.

The blinker or pulsating light-signal consists of a single light-source mechanically occulted. It is controlled by means of a telegraph-key and the code may be rapidly transmitted. The search-light affords a means for signaling great distances, even in the daytime. The light is usually mechanically occulted by a quick-acting shutter, but recently another system has been devised. In the latter the light itself is controlled by means of an electrical shunt across the arc. In this manner the light is dimmed by shunting most of the current, thereby producing the same effect as actually eclipsing the light with a mechanical shutter. By means of the search-light signals are usually visible as far as the limitations of the earth's curvature will permit. By directing the beam against a cloud, signals have been observed at a distance of one hundred miles from the search-light despite intervening elevated land or the curvature of the ocean's surface. By means of small search-lights it is easy to send signals ten miles.

This kind of apparatus has the advantage of being selective; that is, the signals are not visible to persons a few degrees from the direction of the beam. One of the most recent developments has been a special tungsten filament in a gas-filled bulb placed at the focus of a small parabolic mirror. The beam is directed by means of sights and the flashes are obtained by interrupting the current by means of a trigger-switch. The filament is so sensitive that signals may be sent faster than the physiological process of vision will record. With the advent of wireless telegraphy light-signaling for long distances was temporarily eclipsed, but during the recent war it was revived and much development work was prosecuted.

The Ardois system consists of four lamps mounted in a vertical line as high as possible. Each lamp is double, containing a red and a white light, and these lights are controlled from a keyboard. A red light indicates a dot in the Morse code and a white light indicates a dash. The keys are numbered and lettered, so that the system may be operated by any one. Various other systems employing colored lights have been used, but they are necessarily short-range signals. Another example is the semaphore. When used at night, tungsten lamps in reflectors indicate the positions of the arms. The advantage of these signals over the flashing-system is that each signal is complete and easy to follow. The flashing-system is progressive and must be carefully followed in order to obtain the meaning of the dots and dashes.

Smaller signal-lamps using acetylene have been employed in the forestry service and in other activities where a portable device is necessary. In one type, a mixture-tank containing calcium carbide and water is of sufficient capacity for three hours of signaling. A small pilot-light is permitted to burn constantly and the flashes are obtained by operating a key which increases the gas-pressure. The light flares as long as the key is depressed. The range of this apparatus is from ten to twenty miles. An electric lamp supplied from a storage battery has been designed for geodetic operations in mountainous districts where it is desired to send signals as far as one hundred miles. Tests show that this device is a hundred and fifty times more powerful than the ordinary acetylene signal-lamp, and it is thought that with this new electric lamp haze and smoke will seldom prevent observations.

Certain fixed lights are required by law on a vessel at night. When it is under way there must be a white light at the masthead, a starboard green light, a port red light, a white range-light, and a white light at the stern. The masthead light is designed to emit light through a horizontal arc of twenty points of the compass, ten on each side of dead ahead. This light must be visible at a distance of five miles. The port and starboard lights operate through a horizontal arc of twenty points of the compass, the middle of which is dead ahead. They are screened so as not to be visible across the bow and they must be intense enough to be visible two miles ahead. The masthead light is carried on the foremast and the range-light on the mainmast, at an elevation fifteen feet higher than the former. The range-light emits light toward all points of the compass and must be intense enough to be seen at a distance of three miles. The stern light is similar to the masthead, but its light must not be visible forward of the beam. When a vessel is towing another it must display two or three lights in a vertical line with the masthead light and similar to it. The lights are spaced about six feet apart, and two extra ones indicate a short tow and three a long one. A vessel over a hundred and fifty feet long when at anchor is required to display a white light forward and aft, each visible around the entire horizon. These and many other specifications indicate how artificial light informs the mariner and makes for order in shipping. Without artificial light the waterways would be trackless and chaos would reign.

The distress signals of a vessel are rockets, but any burning flame also serves if rockets are unavailable. Fireworks were known many centuries ago and doubtless the possibilities of signaling by means of rockets have long been recognized. An early instance of scientific interest in rockets and their usefulness is that of Benjamin Robins in 1749. While he was witnessing a display of fireworks in London it occurred to him that it would be of interest to measure the height to which the rockets ascended and to determine the ranges at which they were visible. His measurements indicated that the rockets ascended usually to a height of 440 yards, but some of them attained altitudes as high as 615 yards. He then had some special ones made and despatched letters to friends in three different localities, at distances as great as 50 miles, asking them to observe at a certain time, when the rockets were to be sent up in the outskirts of London. Some of these rockets rose to altitudes as great as 600 yards and were distinctly seen by observers 38 miles away. Later he made rockets which ascended as high as 1200 yards and concluded that this was a practical means of signaling. Since that time and especially during the recent war, rockets have served well in signaling messages.

The self-propelled rockets have not been altered in essential features since the remote centuries when the Chinese first used them in celebrations. A cylindrical shell is mounted on a wooden stick and when the powder in the shell burns the hot gases are ejected so violently downward that the reaction drives the shell upward. At a certain point in the air, various signals burst forth, which vary in character and color. One of the advantages of the rocket is that it contains within itself the force of propulsion; that is, no gun is necessary to project it. The illuminating compounds and various details are similar to those of the illuminating shells described in another chapter.

At present the rocket is not scientifically designed to obtain the greatest efficiency of propulsion, but its simplicity in this respect is one of its chief advantages. If the self-propelled rocket becomes the projectile of the future, as some have ventured to predict, much consideration must be given to the design of the orifice through which the gases violently escape in order that the best efficiency of propulsion may be attained. There are other details in which improvements may be made. The combustion products of the black powder which are not gaseous equal about one third the weight of the powder. This represents inefficient propulsion. Furthermore, during recent years much information has been gained pertaining to the air-resistance which can be applied to advantage in designing the form of rockets.

Besides the various rockets, signal-lights have been constructed to be fired from guns and pistols. During the recent war the airman in the dark heights used the pistol signal-light effectively for communication. These devices emitted stars either singly or in succession, and the color of these stars as well as their number and sequence gave significance to the signal. Some of these light-signals were provided with parachutes and were long-burning; that is, light was emitted for a minute or two. There are many variations possible and a great many different kinds of light-signals of this character were used. In the front-line trenches and in advances they were used when telephone service was unavailable. The airman directed artillery fire by means of his pistol-light. Rockets brought aid to the foundered ship or to the life-boats. The signal-tube which burned red, green, or white was held in the hand or laid on the ground and it often told its story. For many years such a device dropped from the rear of the railroad train has kept the following train at a safe distance. A device was tried out in the trenches, during the war, which emitted a flame. This could be varied in color to serve as a signal and the apparatus had sufficient capacity for thirty hours' burning. This could also be used as a weapon, or when reduced in intensity it served as a flash-light.

For many years experiments have been made upon the use of the invisible rays which accompany visible rays. The practicability of signaling with invisible rays depends upon producing them efficiently in sufficient quantity and upon separating them from the visible rays which accompany them. Some successful results were obtained with a 6-volt electric lamp possessing a coiled filament at the focus of a lens three inches in diameter and twelve inches in focal length. This gave a very narrow beam visible only in the neighborhood of the observation post to which the signals were directed. The beam was directed by telescopic sights. During the day a deep red filter was placed over the lamp and the light was invisible to an observer unless he was equipped with a similar red screen to eliminate the daylight. It is said that signals were distinguished at a distance of six miles. By night a screen was used which transmitted only the ultraviolet rays, and the observer's telescope was provided with a fluorescent screen in its focal plane. The ultraviolet rays falling upon this screen were transformed into visible rays by the phenomenon of fluorescence. The range of this device was about six miles. For naval convoys lamps are required to radiate toward all points of the compass. For this purpose a quartz mercury-arc which is rich in ultraviolet rays was surrounded with a chimney which transmitted the ultraviolet rays efficiently and absorbed all visible rays excepting violet light. The lamp appeared a deep violet color at close range, but the faintly visible light which it transmitted was not seen at a distance. A distant observer picks up the invisible ultraviolet "light" by means of a special optical device having a fluorescent screen of barium-platino-cyanide. This device had a range of about four miles.

Light-signals are essential for the operation of railways at night and they have been in use for many years. In this field the significance of light-signals is based almost universally on color. The setting of a switch is indicated by the color of the light that it shows. With the introduction of the semaphore system, in which during the day the position of the arm is significant, colored glasses were placed on the opposite end of the arm in such a manner that a certain colored glass would appear before the light-source for a certain position of the arm. A kerosene flame behind a glass lens was the lamp used, and, for example, red meant "Stop," green counseled "Caution," and clear or white indicated "All clear." For many years the kerosene lamp has been used, but recently the electric filament lamp is being installed to some extent for this purpose. In fact, on one railroad at least, tungsten lamps are used for light-signals by day as well as by night. Three signals--red, green, and white-- are placed in a vertical line and behind each lens are two lamps, one operating at high efficiency and one at low efficiency to insure against the failure of the signal. The normal daylight range is about three thousand feet and under the worst conditions when opposed to direct sunlight, the range is not less than two thousand feet. It is said that these lights are seen more easily than semaphore arms under all circumstances and that they show two or three times as far as the latter during a snow-storm.

The standard colors for light-signals as adopted by the Railway Signal Association are red, yellow, green, blue, purple, and lunar white. These are specified as to the amount of the various spectral colors which they transmit when the light-source is the kerosene flame. Obviously, the colors generally appear different when another illuminant is used. The blue and purple are short-range signals, but the effective range of the best railway signal employing a kerosene flame is only about four miles.

It has been shown that the visibility of point sources of white light in clear atmosphere, for distances up to a mile at least, is proportional to their candle-power and inversely proportional to the square of the distance. Apparently the luminous intensities of signal-lamps required in clear weather in order that they may be visible must be 0.43 candles for one nautical mile, 1.75 candles for two nautical miles, and 11 candles for five nautical miles. From the data available it appears that a red or a white signal-light will be easily visible at a distance in nautical miles equal to the square root of its candle-power in that direction. The range in nautical miles of a green light apparently is proportional to the cube root of the candle-power. Whether or not these relations between the range in miles and the luminous intensity in candles hold for greater distances than those ordinarily encountered has not been determined, but it is interesting to note that the square root of the luminous intensity of the Navesink Light at the entrance to New York Harbor is about 7000. Could this light be seen at a distance of seven thousand miles through ordinary atmosphere?

The most distinctive colored lights are red, yellow, green, and blue. To these white (clear) and purple have been added for signaling-purposes. Yellow is intense, but it may be confused with "white" or clear. Blue and purple as obtained from the present practicable light-sources are of low intensity. This leaves red, green, and clear as the most generally satisfactory signal-lights.

There are numerous other applications, especially indoors. Some of these have been devised for special needs, but there are many others which are general, such as for elevators, telephones, various call systems, and traffic signals. Light has the advantages of being silent and controllable as to position and direction, and of being a visible signal at night. Thus, in another field artificial light has responded to the demands of civilization.

XVI

THE COST OF LIGHT

Artificial light is so superior to natural light in many respects that mankind has acquired the habit of retiring many hours after darkness has fallen, a result of which has brought forth the issue known as "daylight saving." Doubtless, daylight should be used whenever possible, but there are two sides to the question. In the first place, it costs something to bring daylight indoors. The architectural construction of windows and skylights increases the cost of daylight. Light-courts, by sacrificing valuable floor-area, add to the expense. The maintenance of windows and sky lights is an appreciable item. Considering these and other factors, it can be seen that daylight indoors is expensive; and as it is also undependable, a supplementary system of artificial lighting is generally necessary. In fact, it is easy to show in some cases that artificial lighting is cheaper than natural lighting.

The average middle-class home is now lighted artificially for about $15.00 to $25.00 per year, with convenient light-sources which are available at all times. There is no item in the household budget which returns as much satisfaction, comfort, and happiness in proportion to its cost as artificial light. It is an artistic medium of great potentiality, and light in a narrow utilitarian sense is always a by-product of artistic lighting. The insignificant cost of modern lighting may be emphasized in many ways. The interest on the investment in a picture or a vase which cost $25.00 will usually cover the cost of operating any decorative lamp in the home. A great proportion of the investment in personal property in a home is chargeable to an attempt to beautify the surroundings. The interest on only a small portion of this investment will pay for artistic and utilitarian artificial lighting in the home. The cost of washing the windows of the average house may be as great as the cost of artificial lighting and is usually at least a large fraction of the latter. It would become monotonous to cite the various examples of the insignificant cost of artificial light and its high return to the user. The example of the home has been chosen because the reader may easily carry the analysis further. The industries where costs are analyzed are now looking upon adequate and proper lighting as an asset which brings in profits by increasing production, by decreasing spoilage, and by decreasing the liability of accidents.

Inasmuch as daylight saving became an issue during the recent war and is likely to remain a matter of concern, its history is interesting. One of the outstanding differences between primitive and civilized beings is their hours of activities. The former automatically adjusted themselves to daylight, but as civilization advanced, the span of activities began to extend more and more beyond the coming of darkness. Finally in many activities the work-day was extended to twenty-four hours. There can be no insurmountable objection to working at night with a proper arrangement of the periods of work; in fact, the cost of living would be greatly increased if the overhead charges represented by such items as machinery and buildings were allowed to be carried by the decreased products of a shortened period of production. There cannot be any basic objection to artificial lighting, because most factories, for example, may be better illuminated by artificial than by natural light.

Of course, the lag of comfortable temperature behind daylight is responsible to some extent for a natural shifting of the ordinary working-day somewhat behind the sun. The chill of dawn tends to keep mankind in bed and the cheer of artificial light and the period of recreation in the evening tends to keep the civilized races out of bed. There are powerful influences always at work and despite the desirable features of daylight-saving, mankind will always tend to lag. As years go by, doubtless it will be necessary to make the shift again and again. It seems certain that throughout the centuries thoughtful persons have seen the difficulty of rousing man from his warm bed in the early morning and have recognized a simple solution in turning the hands of the clock ahead. Among the earliest advocates of daylight saving during modern times, when it became important enough to be considered as an economic issue, was Benjamin Franklin. In 1784 he wrote a masterful serio-comic essay entitled "An Economical Project" which was published in the Journal of Paris. The article, which appeared in the form of a letter, began thus:

MESSIEURS: You often entertain us with accounts of new discoveries. Permit me to communicate to the public through your paper one that has lately been made by myself and which I conceive may be of great utility.

I was the other evening in a grand company where the new lamp of Messrs. Quinquet and Lange was introduced and much admired for its splendor; but a general inquiry was made whether the oil it consumed was not in exact proportion to the light it afforded, in which case there would be no saving in the use of it. No one present could satisfy us on that point, which all agreed ought to be known, it being a very desirable thing to lessen, if possible, the expense of lighting our apartments, when every other article of family expense was so much augmented. I was pleased to see this general concern for economy, for I love economy exceedingly.

I went home, and to bed, three or four hours after midnight, with my head full of the subject. An accidental sudden noise waked me about 6 in the morning, when I was surprised to find my room filled with light, and I imagined at first that a number of those lamps had been brought into it; but, rubbing my eyes, I perceived the light came in at the windows. I got up and looked out to see what might be the occasion of it, when I saw the sun just rising above the horizon, from whence he poured his rays plentifully into my chamber, my domestic having negligently omitted the preceding evening to close the shutters.

I looked at my watch, which goes very well, and found that it was but 6 o'clock; and, still thinking it something extraordinary that the sun should rise so early, I looked into the almanac, where I found it to be the hour given for his rising on that day. I looked forward, too, and found he was to rise still earlier every day till toward the end of June, and that at no time in the year he retarded his rising so long as till 8 o'clock.

Your readers who, with me, have never seen any signs of sunshine before noon, and seldom regard the astronomical part of the almanac, will be as much astonished as I was when they hear of his rising so early, and especially when I assure them that he gives light as soon as he rises. I am convinced of this. I am certain of my fact. One cannot be more certain of any fact. I saw it with my own eyes. And, having repeated this observation the three following mornings, I found always precisely the same result.

He then continues in the same vein to show that learned persons did not believe him and to point out the difficulties which the pioneer encounters. He brought out the vital point by showing that if he had not been awakened so early he would have slept six hours longer by the light of the sun and in exchange he would have lived six hours the following night by candle-light. He then mustered "the little arithmetic" he was master of and made some serious computations. He assumed as the basis of his computations that a hundred thousand families lived in Paris and each used a half-pound of candles nightly. He showed that between March 20th and September 20th, 64,000,000 pounds of wax and tallow could be saved, which was equivalent to $18,000,000.

After these serious computations he amusingly proposed the means for enforcing the daylight saving. Obviously, it was necessary to arouse the sluggards and his proposals included the use of cannons and bells. Besides, he proposed that each family be restricted to one pound of candles per week, that coaches would not be allowed to pass after sunset except those of physicians, etc., and that a tax be placed upon every window which had shutters. His closing paragraph was as follows:

For the great benefit of this discovery, thus freely communicated and bestowed by me on the public, I demand neither place, pension, exclusive privilege, nor any other regard whatever. I expect only to have the honor of it. And yet I know there are little, envious minds who will, as usual, deny me this and say that my invention was known to the ancients, and perhaps they may bring passages out of the old books in proof of it. I will not dispute with these people that the ancients knew not the sun would rise at certain hours; they possibly had, as we have, almanacs that predicted it; but it does not follow thence that they knew he gave light as soon as he rose. That is what I claim as my discovery. If the ancients knew it, it might have been long since forgotten; for it certainly was unknown to the moderns, at least to the Parisians, which to prove I need use but one plain simple argument. They are as well instructed, judicious and prudent a people as exist anywhere in the world, all professing, like myself, to be lovers of economy, and, for the many heavy taxes required from them by the necessities of the State have surely an abundant reason to be economical. I say it is impossible that so sensible a people, under such circumstances, should have lived so long by the smoky, unwholesome and enormously expensive light of candles, if they had really known that they might have had as much pure light of the sun for nothing.

Franklin's amusing letter had a serious aim, for in 1784 family expenses were much augmented and adequate lighting by means of candles was very costly in those days. However, conditions have changed enormously in the past hundred and thirty-five years. A great proportion of the population lives in the darker cities. The wheels of progress must be kept going continuously in order to curb the cost of living, which is constantly mounting higher owing to the addition of conveniences and luxuries. Furthermore, the cost of light has so diminished that it is not only a minor factor at present but in many cases is actually paying dividends in commerce and industry. It is paying dividends of another kind in the social and educational aspects of the home, library, church, and art museum. Daylight saving has much to commend it, but the cost of daylight and the value of artificial light are important considerations.

The cost of fuels for lighting purposes cannot be thoroughly compared throughout a span of years without regard to the fluctuating purchasing power of money, which would be too involved for consideration here. However, it is interesting to make a brief survey throughout the past century. From 1800 until 1845 whale-oil sold for about $.80 per gallon, but after this period it increased in value, owing apparently to its growing scarcity, until it reached a price of $1.75 per gallon in 1855. Fortunately, petroleum was discovered about this time, so that the oil-lamp did not become a luxury. From 1800 to 1850 tallow-candles sold at approximately 20 cents a pound. There being six candles to the pound, and inasmuch as each candle burned about seven hours, the light from a candle cost about 1/2 cent per hour. From 1850 to 1875 tallow-candles sold at an average price of approximately 25 cents a pound. It may be interesting to know that a large match emits about as much light as a burning candle and a so-called safety match about one third as much.

A candle-hour is the total amount of light emitted by a standard candle in one hour, and candle-hours in any case are obtained by multiplying the candle-power of the source by the hours of burning. In a similar manner, lumens output multiplied by hours of operation give the lumen-hours. A standard candle may be considered to emit an amount of light approximately equal to 10 lumens. A wax-candle will emit about as much light as a sperm candle but will consume about 10 per cent. less weight of material. A tallow candle will emit about the same amount of light with a consumption about 50 per cent. greater. The tallow-candle has disappeared from use.

With the appearance of kerosene distilled from petroleum the camphene lamp came into use. The kerosene cost about 80 cents per gallon during the first few years of its introduction. The price of kerosene averaged about 55 cents a gallon between 1865 and 1875. During the next decade it dropped to about 22 cents a gallon and between 1885 and 1895 it sold as low as 13 cents.

Artificial gas in 1865 sold approximately at $2.50 per thousand cubic feet; between 1875 and 1885 at $2.00; between 1885 and 1895 at $1.50.

The combined effect of decreasing cost of fuel or electrical energy for light-sources and of the great improvements in light-production gave to the householder, for example, a constantly increasing amount of light for the same expenditure. For example, the family which a century ago spent two or three hours in the light of a single candle now enjoys many times more light in the same room for the same price. It is interesting to trace the influence of this greatly diminishing cost of light in the home. For the sake of simplicity the light of a candle will be retained as the unit and the cost of light for the home will be considered to remain approximately the same throughout the period to be considered. In fact, the amount of money that an average householder spends for lighting has remained fairly constant throughout the past century, but he has enjoyed a longer period of artificial light and a greater amount of light as the years advanced. The following is a table of approximate values which shows the lighting obtainable for $20.00 per year throughout the past century exclusive of electricity:

Year	Hours per night	Equivalent light in candles	Candle-hours per night	Candle-hours per year
1800	3	5	15	5,500
1850	3	8	24	8,700
1860	3	11	33	12,000
1870	3	22	66	24,000
1880	3.5	36	126	46,000
1890	4	50	200	73,000
1900	5	154	770	280,000

It is seen from the foregoing that in a century the candle-equivalent obtainable for the same cost to the householder increased at least thirty times, while the hours during which this light is used have nearly doubled. In other words, in the nineteenth century the candle-hours obtainable for $20.00 per year increased about fifty times. Stated in another manner, the cost of light at the end of the century was about one fiftieth that of candle light at the beginning of the century. One authority in computing the expense of lighting to the householder in a large city of this country has stated that

coincident with an increase of 1700 per cent. in the amount of night lighting of an American family, in average circumstances, using gas for light, there has come a reduction in the cost of the year's lighting of 34 per cent. or approximately $7.50 per year; and that the cost of lighting per unit of light-- the candle-hour--is now but 2.8 per cent. of what it was in the first half of the nineteenth century. No other necessity of household use has been so cheapened and improved during the last century.

In general, the light-user has taken advantage of the decrease by increasing the amount of light used and the period during which it is used. In this manner the greatly diminished cost of light has been a marked sociological and economic influence.

After Murdock made his first installation of gas-lighting in an industrial plant early in the nineteenth century, he published a comparison of the expense of operation with that of candle-lighting. He arrived at the costs of light equivalent to 1000 candle-hours as follows:

1000 candle-hours Gas-lighting at a rate of two hours per day $1.95 " " " " " three " " " 1.40 Candle-lighting 6.50

It is seen that the longer hours of burning reduce the cost of gas-lighting by reducing the percentage of overhead charges. There are no such factors in lighting by candles because the whole "installation" is consumed. This is an early example of which an authentic record is available. At the present time a certain amount of light obtained for $1.00 with efficient tungsten filament lamps, costs $2.00 if obtained from kerosene flames and about $50.00 if obtained by burning candles.

In order to obtain the cost of an equivalent amount of light throughout the past century a great many factors must be considered. Obviously, the results obtained by various persons will differ owing to the unavoidable factor of judgment; however, the following list of approximate values will at least indicate the trend of the price of light throughout the century or more of rapid developments in light-production. A fair average of the retail values of fuels and of electrical energy and an average luminous efficiency of the light-sources involved have been used in making the computations. The figures apply particularly to this country.

TABLE SHOWING THE APPROXIMATE TOTAL COST OF 1000 CANDLE-HOURS FOR VARIOUS PERIODS

Per 1000 candle-hours 1800 to 1850, sperm-oil $2.40 tallow candle 5.00 1850 to 1865, kerosene 1.65 tallow candle 6.85 1865 to 1875, kerosene .75 tallow candle 6.25 gas, open-flame .90 1875 to 1885, kerosene .25 gas, open-flame .60 1885 to 1895, kerosene .15 gas, open-flame .40 1895 to 1915, gas mantle .07 carbon filament .38 metallized filament .28 tungsten filament (vacuum) .12 tungsten filament (gas-filled) .07

In these days the cost of living has claimed considerable attention and it is interesting to compare that of lighting. In the following table the price of food and of electric lighting are compared for twenty years preceding the recent war. The great disturbance due to the war is thereby eliminated from consideration, but it should be noted that since 1914 the price of food has greatly increased but that of electric lighting has not changed materially. The cost of each commodity is taken as one hundred units for the year 1894 but, of course, the actual cost of living for the householder is perhaps a hundred times greater than the cost of electric lighting.

Year Food Electric lighting 1894 100 100 1896 80 92 1898 92 90 1900 100 85 1902 113 77 1904 110 77 1906 115 57 1908 128 30 1910 138 28 1912 144 23 1914 145 17

One feature of electric lighting which puzzles the consumer and which gives the politicians an opportunity for crying "discrimination" and "injustice" at the public-service company is the great variation in rates. There is no discrimination or injustice when the householder, for example, must pay more for his lighting than a factory pays. The rates are not only affected by "demand" but by the period in which the demand comes. Residence lighting is chiefly confined to certain hours from 5 to 9 P. M. and there is a great "peak" of demand at this time. The central-stations must have equipment available for this short-time demand and much of the capacity of the equipment is unused during the remainder of the day. The factory which uses electricity throughout the day or night or both is helping to keep the central-station operating efficiently. The equipment necessary to supply electricity to the factory is operating long hours. Not only is this overhead charge much less for factories and many other consumers than for the householder, but the expense of accounting, of reading meters, etc., is about the same for all classes of consumers. Therefore, this is an appreciable item on the bill of the small consumer.

Doubtless, the public does not realize that the enormous decrease in the cost of lighting during the past century is due largely to the fact that the lighting industry has grown large. Increased production is responsible for some of this decrease and science for much of it. The latter, having been called to the aid of the manufacturers, who are better able by virtue of their magnitude to spend time and resources upon scientific developments, has responded with many improvements which have increased the efficiency of light-production. Some figures of the Census Bureau may be of interest. These are given for 1914 in order that the abnormal conditions due to the recent war may be avoided. The figures pertaining to the manufacture of gas for sale which do not include private plants are as follows for the year 1914 for this country:

Number of establishments 1,284 Capital $1,252,421,584 Value of products (gas, coke, tar, etc.) $220,237,790 Cost of materials $76,779,288 Value added by manufacture $143,458,502 Value of gas $175,065,920 Coal used (tons) 6,116,672 Coke used (tons) 964,851 Oil used (gallons) 715,418,623 Length of gas mains (miles) 58,727 Manufactured products sold Total gas (cubic feet) 203,639,260,000 Straight coal gas (cubic feet) 10,509,946,000 Carbureted water gas (cubic feet) 90,017,725,000 Mixed coal- and water-gas (cubic feet) 86,281,339,000 Oil gas (cubic feet) 16,512,274,000 Acetylene (cubic feet) 136,564,000 Other gas, chiefly gasolene (cubic feet) 181,412,000 Coke (bushels) 114,091,753 Tar (gallons) 125,938,607 Ammonia liquors (gallons) 50,737,762 Ammonia, sulphate (pounds) 6,216,618

Of course, only a small fraction of the total gas manufactured is used for lighting.

According to the U. S. Geological Survey, the quantities of gas sold in this country in the year 1917 were as follows:

Coal-gas 42,927,728,000 cubic feet Water-gas 153,457,318,000 " " Oil-gas 14,739,508,000 " " Byproduct gas 131,026,575,000 " " Natural gas 795,110,376,000 " "

In 1914 there were 38,705,496 barrels (each fifty gallons) of illuminating oils refined in this country and the value was $96,806,452. About half of this quantity was exported. In 1914 the value of all candles manufactured in this country was about $2,000,000, which was about half that of the candles manufactured in 1909 and in 1904. In 1914 the value of the matches manufactured in this country was $12,556,000. This has increased steadily from $429,000 in 1849. In 1914 the glass industries in this country made 7,000,000 lamps, 70,000,000 chimneys, 16,300,000 lantern globes,

24,000,000 shades, globes, and other gas goods. Many millions of other lighting accessories were made, but unfortunately they are not classified.

Some figures pertaining to public electric light and power stations of the United States for the years 1907 and 1917 are as follows:

1917 1907 Number of establishments 6,541 4,714 Commercial 4,224 3,462 Municipal 2,317 1,562 Income $526,886,408 $175,642,338 Total horse-power of plants 12,857,998 4,098,188 Steam engines 8,389,389 2,693,273 Internal combustion engines 217,186 55,828 Water-wheels 4,251,423 1,349,087 Kilowatt capacity of generators 9,001,872 2,709,225 Output in millions of kilowatt-hours 25,438 5,863 Motors served (horse-power) 9,216,323 1,649,026 Electric-arc street-lamps served 256,838 .... Electric-filament street-lamps served 1,389,382 ....

In general, there is a large increase in the various items during the decade represented. The output of the central stations doubled in the five years from 1907 to 1912, and doubled again in the next five years from 1912 to 1917. Street lamps were not reported in 1907, but in 1912 there were 348,643 arc-lamps served by the public companies. The number of arc-lamps decreased to 256,838 in 1917. On the other hand, there were 681,957 electric filament street lamps served in 1912, which doubled in number to 1,389,382 in 1917. The cost of construction and equipment of these central stations totaled more than $3,000,000,000 in 1917.

Although there is no immediate prospect of the failure of the coal and oil supplies, exhaustion is surely approaching. And as the supplies of fuel for the production of gas and electricity diminish, the cost of lighting may advance. The total amount of oil available in the known oil-fields of this country at the present time has been estimated by various experts between 5,000,000,000 and 20,000,000,000 barrels, the best estimate being about 7,000,000,000. The annual consumption is now about 400,000,000 barrels. These figures do not take into account the oil which may be distilled from the rich shale deposits. Apparently this source will yield a hundred billion barrels of oil. In a similar manner the coal-supply is diminishing and the consumption is increasing. In 1918 more than a half-billion tons of coal were shipped from the mines. The production of natural gas perhaps has reached its peak, and, owing to its relation to the coal and oil deposits, its supply is limited.

Although only a fraction of the total production of gas, oil, and coal is used in lighting, the limited supply of these products emphasizes the desirability of developing the enormous water-power resources of this country. The present generation will not be hard pressed by the diminution of the supply of gas, oil, and coal, but it can profit by encouraging and even demanding the development of water-power. Furthermore, it is an obligation to succeeding generations to harness the rivers and even the tides and waves in order that the other resources will be conserved as long as possible. Science will continue to produce more efficient light-sources, but the cost of light finally is dependent upon the cost of the energy supplied to these lamps. At the present time water-power is the anchor to the windward.

XVII

LIGHT AND SAFETY

It is established that outdoors life and property are at night safer under adequate lighting than they are under inadequate lighting. Police departments in the large cities will testify that street-lighting is a powerful ally and that crime is fostered by darkness. But in reckoning the cost of street-lighting to-day how many take into account the value of safety to life and property and the saving occasioned by the reduction in the police-force necessary to patrol the cities and towns? Owing to the necessity of darkening the streets in order to reduce the hazards of air-raids, London experienced a great increase in accidents on the streets, which demonstrated the practical value of street-lighting from the standpoint of accident prevention.

During the war, when dastardly traitors and agents of the enemy were striking at industry, the value of lighting was further recognized by the industries, with the result that flood-lighting was installed to protect them. By common consent this new phase was termed "protective lighting." Soon after the entrance of this country into the recent war, the U. S. Military Intelligence established a Section of Plant Protection which had thirty-three district offices during the war and gave attention to thirty-five thousand industrial plants engaged in production of war materials. Protective lighting was early recognized by this section as a very potential agency for defense, and extensive use was made of it. For example, Edmund Leigh, chief of the section, in discussing the value of outdoor lighting stated:

An illustration of our work in this connection is the case of an $80,000,000 powder plant of recent construction. We arranged to have all wires buried. In addition to the ordinary lighting on an adjacent hill there is a large searchlight which will command any part of the buildings and grounds. Every three hundred yards there is a watch-tower with a searchlight on top. These searchlights are for use only in emergency. Each tower has a telephone service, one connected with the other. The men in the towers have a view of the building exteriors, which are all well lighted, and the men in the buildings look across the yard to the lighted fence line and so get a silhouette of persons or objects in between. The most vital parts of the buildings are surrounded by three fences. In the near-by woods the underbrush has been cleared out and destroyed. The trunks and limbs of trees have been whitewashed. No one can walk among these trees or between the trees and the plant without being seen in silhouette.... I say flatly that I know nothing that is so potential for good defense as good illumination and at the same time so little understood.

Without such protective lighting an army of men would have been required to insure the safety of this one vital plant; still it is obvious that the cost of the protective lighting was an insignificant part of the value of the plant which it insured against damage and destruction.

The United States participated for nineteen months in the recent war and during that time about 400,000 casualties were suffered by its forces. This was at the rate of about 250,000 per year, which included casualties in battle, at sea, and from sickness, wounds, and accidents. Every one has felt the magnitude of this rate of casualties because either his home or that of a friend was blighted by one or more of these tragedies in the nineteen months. However, R. E. Simpson of the Travelers Insurance Company has stated that:

During a one-year period in this country the number of accidents due to inadequate or improper lighting exceeds the yearly rate of our war casualties.

This is a startling comparison, which emphasizes a phase of lighting that has long been recognized by experts but has been generally ignored by the industries and by the public. The condition doubtless is due largely to a lag in the proper utilization of artificial lighting behind the rapid increase in congestion in the industries and in public places.

Accident prevention is an important phase of modern life which must receive more attention. From published statistics and conservative estimates it has been concluded that there are approximately 25,000 persons killed or permanently disabled, 500,000 seriously injured, and 1,000,000 slightly injured each year in this country. Translating these figures by means of the accident severity rates, Mr. Simpson has found that there is a total of 180,000,000 days of time lost per year. This is equivalent to the loss of services of 600,000 men for a full year of 300 work-days. This loss is distributed over the entire country and consequently its magnitude is not demonstrated excepting by statistics. Of course, the causes of the accidents are numerous, but, among the means of prevention, proper lighting is important.

According to some authorities at least 18 per cent. of these accidents are due to defects in lighting. On this basis the services of 108,000 men as producers and wage-earners are continually lost at the present time because the lighting is not sufficient or proper for the safety of workers. If the full year's labor of 108,000 men could be applied to the mining of coal, 130,000,000 million tons of coal would be added to the yearly output; and only 10,000 tons would be necessary to supply adequate lighting for this army of men working for a full year for ten hours each day.

Statistics obtained under the British workmen's compensation system show that 25 per cent. of the accidents were caused by inadequate lighting of industrial plants.

Much has been said and actually done regarding the saving of fuel by curtailing lighting, but the saving may easily be converted into a great loss. For example, a 25-watt electric lamp may be operated ten hours a day for a whole year at the expense of one eighth of a ton of coal. Suppose this lamp to be over a stairway or at any vital point and that by extinguishing it there occurs a single accident which involves the loss of only one day's work on the part of the worker. If this one day's time could have produced coal, there would have been enough coal mined in the ten hours to operate the lamp for thirty-two years. The insignificant cost of lighting is also shown by the distribution of the consumption of fuel for heating, cooking, and lighting in the home. Of the total amount of fuel consumed in the home for these purposes, 87 per cent. is for heating, 11 per cent. for cooking and 2 per cent. for lighting. The amount of coal used for lighting purposes in general is about 2.5 per cent. of the total consumption of coal, so it is seen that the curtailment of lighting at best cannot save much fuel; and it may actually result in a great economic loss. By replacing inefficient lamps and accessories with efficient lighting-equipment and by washing windows and artificial lighting devices, a real saving can be realized.

Improper lighting may be as productive of accidents as inadequate lighting, and throughout the industries and upon the streets the misuse of light is in evidence. The blinding effect of a brilliant light-source is easily proved by looking at the sun. After a few moments great discomfort is experienced, and on looking away from this brilliant source the eyes are temporarily blinded by the after-images. When this happens in a factory as the result of gazing into an unshielded light-source, the workman may be injured by moving machinery, by stumbling over objects, and in many other ways. Unshaded light-sources are too prevalent in the industries. Improper lighting is likely to cause deep shadows wherein many dangers may be hidden. On the street the glare from automobile head-lamps is very prevalent and nearly everybody may testify from experience to the dangers of glare. Even the glaring locomotive head-lamp has been responsible for many casualties.

Unfortunately, natural lighting outdoors has not been under the control of man and he has accepted it as it is. The sky is a harmless source of light when viewed outdoors and the sun is in such a position that it is usually easy to avoid looking at it. It is so intensely glaring that man unconsciously avoids looking directly at it. These conditions are responsible to an extent for man's indifference and even ignorance of the rudiments of safe lighting. When he has artificial light, over which he may exercise control, he either ignores it or owing to the less striking glare he misuses it and his eyesight without realizing it. A great deal of eye-strain and permanent eye trouble arises from the abuse of the eyes by improper lighting. For example, near-sightedness is often due to inadequate illumination, which makes it necessary for the eyes to be near the work or the reading-page. Improper or inadequate lighting especially influences eyes that are immature in growth and in function, and it has been shown that with improvements in lighting the percentage of short-sightedness has decreased in the schools. Furthermore, it has been shown that where no particular attention has been given to lighting and vision, the percentage of short-sightedness has increased with the grade. There are twenty million school children in this country whose future eyesight is in the hands of those who have jurisdiction over lighting and vision. There are more than a hundred million persons in this country whose eyes are daily subjected to improper lighting-conditions, either through their own indifference or through the negligence of others.

Of a certain group of 91,000 purely industrial accidents in the year 1910, Mr. Simpson has stated that 23.8 per cent. were due, directly or indirectly, to the lack of proper illumination. These may be further divided into two approximately equal groups, one of which comprises the accidents due to inadequate illumination and the other to those toward which improper lighting was a contributing cause. The seasonal variation of these accidents is given in the following table, both for those due directly or indirectly to inadequate and improper lighting and those due to other causes.

SEASONAL DISTRIBUTION OF INDUSTRIAL ACCIDENTS DUE TO LIGHTING CONDITIONS AND TO OTHER CAUSES

Percentage due to Lighting conditions Other causes

July 4.8 5.9 August 5.2 6.2 September 6.1 6.9 October 8.6 8.5 November 10.9 10.5 December 15.6 12.2 January 16.1 11.9 February 10.0 10.5 March 7.6 8.8 April 6.1 6.9 May 5.2 5.8 June 3.8 5.9

The figures in one column have no direct relation to those in the other; that is, each column must be considered by itself. It is seen from the foregoing that about half the number of the accidents due to poor illumination occurred in the months of November, December, January, and February. These are the months of inadequate illumination unless artificial lighting has been given special attention. The same general type of seasonal distribution of accidents due to other causes is seen to exist but not so prominently. The greatest monthly rate of accidents during the winter season is nearly four times the minimum monthly rate during the summer for those accidents due to lighting conditions. This ratio reduces to about twice in the case of accidents due to other causes. Looking at the data from another angle, it may be considered that the likelihood of an accident being caused by lighting conditions is about twice as great in any of the four "winter" months as in any of the remaining eight months. Doubtless, this may be explained largely upon the basis of morale. The winter months are more dreary than those of summer and the workman's general outlook is different in winter than in summer. In the former season he goes back and forth to work in the dark, or at best, in the cold twilight. He is not only more depressed but he is clumsier in his heavier clothing. If the enervating influence of these factors is combined with a greater clumsiness due to cold and perhaps to colds, it is not difficult to account for this type of seasonal distribution of accidents. A study of the accidents of 1917 indicated that 13 per cent. occurred between 5 and 6 P. M. when artificial lighting is generally in use to help out the failing daylight. Only 7.3 per cent. occurred between 12 M. and 1 P. M.

There is another aspect of the subject which deals particularly with the safety of the light-source or method of lighting. As each innovation in lighting appeared during the past century there immediately arose the question of safety. The fire-hazard of open flames received attention in early days, and when gas-lighting appeared it was condemned as a poison and an explosive. Mineral-oil lamps introduced the danger of explosions of the vapors produced by evaporation. When electric lighting appeared it was investigated thoroughly. The result of all this has been an effort to make lamps and methods safe. Insurance companies have the relative safety of these systems established to their satisfaction and to-day little fire-hazard is attached to the present modes of general lighting if proper precautions have been taken.

When electric lighting was first introduced the public looked upon electricity as dangerous and naturally many questions pertaining to hazards arose. The distribution of electricity has been so highly perfected that little is heard of the hazards which were so magnified in the early years. Data gathered between 1884 and 1889 showed that about 13,000 fires took place in a certain district. Of these, 42 were attributed to electric wires; 22 times as many to breakage and explosion of kerosene lamps; and ten times as many through carelessness with matches. These figures cannot be taken at their face value because of the absence of data showing the relative amount of electric and kerosene lighting; nevertheless they are interesting because they represent the early period.

There are industries where unusual care must be exercised in regard to the lighting. In certain chemical industries no lamps are used excepting the incandescent lamp and this is enclosed in an air-tight glass globe. Even a public-service gas company cautions its employees and patrons thus: "Do not look for a gas-leak with a naked light! Use electric light." The coal-mine offers an interesting example of the precautions necessary because the same type of problems are found in it as in industries in general, with the additional difficulties attending the presence or possible presence of explosive gas. The surroundings in a coal-mine reflect a small percentage of the light, so that much light is wasted unless the walls are whitewashed. This is a practical method for increasing safety in coal-mines. However, the most dangerous feature is the light-source itself. According to the Bureau of Mines during the years 1916 and 1917 about 60 per cent. of the fatalities due to gas and coal-dust explosions were directly traceable to the use of defective safety lamps and to open flames.

In the early days of coal-mining it was found that the flame of a candle occasionally caused explosions in the mines. It was also found that sparks of flint and steel would not readily ignite the gas or coal-dust and this primitive device was used as a light-source. Of course, statistics are unavailable concerning the casualties in coal-mines throughout the past centuries, but with the accidents not uncommon in this scientific age, with its elaborate organizations striving to stamp out such casualties, there is good reason to believe that previous to a century or two ago the risks of coal-mining must have been great. Open flames have been widely used in this industry, but there has always been the risk of the presence or the appearance of gas or explosive dust.

The early open-flame lamps not only were sources of danger but their feeble varying intensity caused serious damage to the eyesight of miners. This factor is always present in inadequate and improper lighting, but its influence is noticeable in coal-mining in the nervous disease affecting the eyes which is known as nystagmus. The symptoms of the disease are inability to see at night and the dazzling effect of ordinary lamps. Finally objects appear to the sufferer to dance about and his vision is generally very much disturbed.

The oil-lamps used in coal-mining have a luminous intensity equivalent to about one to four candles, but owing to the atmospheric conditions in the mines a flame does not burn as brightly as in the fresh air. The possibility of explosion due to the open flame was eliminated by surrounding it with a metal gauze. Davy was the inventor of this device and his safety lamp introduced about a hundred years ago has been a boon to the coal-miner. Various improvements have been devised, but Davy's lamp contained the essentials of a safety device. The flame is surrounded by a cylinder of metal gauze which by forming a much cooler boundary prevents the mine-gas from becoming heated locally by the lamp flame to a sufficient temperature to ignite and consequently to explode. This device not only keeps the flame from igniting the gas but it also serves as an indicator of the amount of gas present, by the variation in the size and appearance of the tip of the flame. However, the gauze reduces the luminous output, and as it accumulates soot and dust the light is greatly diminished. One of these lamps is about as luminous as a candle, initially, but its intensity is often reduced by accumulations upon the gauze to only one fifth of the initial value.

The acetylene lamp is the best open-flame light-source available to the miner, for several reasons. It is of a higher candle-power than the others and as it is a burning gas, there is not the danger of flying sparks as in the case of burning wicks. The greater intensity of illumination affords a greater safety to the miner by enabling him to detect loose rock which may be ready to fall upon him. However, this lamp may be a source of danger, owing to the fact that it will burn more brilliantly in a vitiated atmosphere than other flame-lamps. Another disadvantage is the possibility of calcium carbide accidentally spilt coming in contact with water and thereby causing the generation of acetylene gas. If this is produced in the mine in sufficient quantities it is a danger which may not be suspected. If ignited it will explode and may also cause severe burns.

The electric lamp, being an enclosed light-source capable of being subdivided and fed by a small portable battery, early gave promise of solving the problem of a safe mine-lamp of adequate candle-power. Much ingenuity has been applied to the development of a portable electric safety mine-lamp, and several such lamps are now approved by the Bureau of Mines. Two general types are being manufactured, the cap outfit and the hand outfit. They consist essentially of a lamp in a reflector whose aperture is closed with a sheet or a lens of clear glass. The battery may be of the "dry" or "storage" type and in the case of the cap outfit the battery is carried on the back. The specifications for these lamps demand that a luminous intensity averaging at least 0.4 candle be maintained throughout twelve consecutive hours of operation. At no time during this period shall the output of light fall below 1.25 lumens for a cap-lamp and below 3 lumens for a hand-lamp. Inasmuch as these are equipped with reflectors, the specifications insist that a circle of light at least seven feet in diameter shall be cast on a wall twenty inches away. It appears that a portable lamp is an economic necessity in the coal-mines, on account of the expense, inconvenience, and possible dangers introduced by distribution systems such as are used in most places.

Although the major defects in lighting are due to absence of light in dangerous places, to glare, and to other factors of improper lighting, there are many minor details which may contribute to safety. For example, low lamps are useful in making steps in theaters and in other places, in drawing attention to entrances of elevators, in lighting the aisles of Pullman cars, under hand-rails on stairways, and in many other vital places. A study of accidents indicates that simple expedients are effective preventives.

XVIII

THE COST OF LIVING

A comparison of the civilization of the present with that of a century ago reveals a startling difference in the standards of living. To-day mankind enjoys conveniences and luxuries that were undreamed of by the past generations. For example, a certain town in Iowa, a score of years ago, was appraised for a bond-issue and it was necessary to extend its limits considerably in order to include a valuation of one half million dollars required by the underwriters. On a summer's evening at the present time a thousand "pleasure" automobiles may be found parked along its streets and these exceed in valuation that of the entire town only twenty years ago and equal it to-day. There are economists who would argue that the automobile has paid for itself by its usefulness, but the fact still exists that a great amount of labor has been diverted from producing food, clothing, and fuel to the production of "pleasure" automobiles. And this is the case with many other conveniences and luxuries. It is admitted that mankind deserves these refinements of modern civilization, but he must expect the cost of living to increase unless counteracting measures are taken.

The economics of the increasing cost of living and the analysis of the relations of necessities, conveniences, and luxuries are too complex to be thoroughly discussed here. In fact, the most expert economists would disagree on many points. However, it is certain that the cost of living has steadily increased during the past century and it is reasonably certain that the standards of the present civilization are responsible for some if not all of the increase. Increased production is an anchor to the windward. It may drag and give way to some extent, but it will always oppose the course of the cost of living.

When the first industrial plant was lighted by gas, early in the nineteenth century, the aim was merely to reinforce daylight toward the end of the day. Continuous operation of industrial plants was not practised in those days, excepting in a very few cases where it was essential. To-day some industries operate continuously, but most of them do not. In the latter case the consumer pays more for the product because the percentage of fixed or overhead charge is greater. Investment in ground, buildings, and equipment exacts its toll continuously and it is obvious that three successive shifts producing three times as much as a single day shift, or as much as a trebled day shift, will produce the less costly product. In the former case the fixed charge is distributed over the production of continuous operation, but in the latter case the production of a single day shift assumes the entire burden. Of course, there are many factors which enter into such a consideration and an important one is the desirability of working at night. It is not the intention to touch upon the psychological and sociological aspects but merely to look coldly upon the facts pertaining to artificial light and production.

In the first place, it has been proved that in factories proper lighting as obtained by artificial means is generally more satisfactory than the natural lighting. Of course, a narrow building with windows on two sides or a one-story building with a saw-tooth roof of best design may be adequately illuminated by natural light, but these buildings are the exception and they will grow rarer as industrial districts become more congested. Artificial light may be controlled so that light of a satisfactory quality is properly directed and diffused. Sufficient intensities of illumination may be obtained and the failure of artificial light is a remote possibility as compared with the daily failure of natural light. With increasing cost of ground space, factories are built of several stories and with less space given to light courts, with the result that the ratio of window area to that of the floor is reduced. These tendencies militate against satisfactory daylighting. In the smoky congested industrial districts the period of effective daylight is gradually diminishing and artificial lighting is always essential at least as a reinforcement for daylight. It has been proved that proper artificial lighting--and there is no excuse for improper artificial lighting--is superior to most interior daylighting conditions.

Although it is difficult to present figures in a brief discussion of this character, it may be stated that, in general, the cost of adequate artificial light is about 2 per cent. of the pay-roll of the workers; about 10 per cent. of the rental charges; and only a fraction of 1 per cent. of the cost of the manufactured products. These figures vary considerably, but they represent conservative average estimates. From these it is seen that artificial lighting is a small factor in adding to the cost of the product. But does artificial lighting add to the cost of a product? Many examples could be cited to prove that proper artificial lighting may be responsible for an actual reduction in the cost of the product.

In a certain plant it was determined that the workmen each lost an appreciable part of an hour per day because of inadequate lighting. A properly designed and maintained lighting-system was installed and the saving in the wages previously lost, more than covered the operating-expense of the artificial lighting. Besides really costing the manufacturer less than nothing, the new artificial lighting system was responsible for better products, decreased spoilage, minimized accidents, and generally elevated spirits of the workmen. In some cases it is only necessary to save one minute per hour per workman to offset entirely the cost of lighting. The foregoing and many other examples illustrate the insignificance of the cost of lighting.

The effectiveness of artificial lighting in reducing the cost of living is easily demonstrated by comparing the output of a factory operating on one and two shifts per day respectively. In a well-lighted factory which operated day and night shifts, the cost of adequate lighting was 7 cents per square foot per year. If this factory, operating only in the daytime, were to maintain the same output, it would be necessary to double its size. In order to show the economic value of artificial lighting it is only necessary to compare the cost of lighting with the rental charge of the addition and of its equipment. A fair rental value for plant and equipment is 50 cents per square foot per year; but of course this varies considerably, depending upon the type of plant and the character of the equipment. An investigation showed that this value varies usually between 30 to 70 cents per square foot per year. Using the mean value, 50 cents, it is seen that the rental charge is about seven times the cost of lighting. Furthermore, there is a saving of 43 cents per square foot per year during the night operation by operating the night shift. Of course, this is not strictly true because a depreciation of machinery during the night shift should be allowed for. These fixed charges would average slightly more than half as much in the case of the two-shift factory as in the case of the same output from a factory twice as large but operating only a day shift. Incidentally, the two-shift factory need not be a hardship for the workers, for, if the eight-hour shifts are properly arranged, the worker on the night shift may be in bed by midnight and the objection to a disturbance of ordinary hours of sleep is virtually eliminated.

In a discussion of light and safety presented in another chapter the startling industrial losses due to accidents are shown to be due partially to inadequate or improper lighting. About one fourth of the total number of accidents may be charged to defective lighting. The consumer bears the burden of the support of an unproducing army of idle men. According to some experts an average of about 150,000 men are continuously idle in this country owing to inadequate and improper lighting.

This is an appreciable factor in the cost of living, but the greatest effectiveness of artificial lighting in curtailing costs is to be found in reducing the fixed charges borne by the product through the operation of two shifts and by directly increasing production owing to improved lighting. The standard of artificial-lighting intensity possessed by the average person at the present time is an inheritance from the past. In those days when artificial light was much more costly than at present the tendency naturally was to use just as little light as necessary. That attitude could not have been severely criticized in those early days of artificial lighting, but it is inexcusable to-day. Eyesight and greater safety from accidents are in themselves valuable enough to warrant adequate lighting, but besides these there is the appeal of increased production.

Outdoors on a clear summer day at noon the intensity of daylight illumination at the earth's surface is about 10,000 foot-candles; in other words, it is equal to the illumination on a surface produced by a light-source equivalent to 10,000 candles at a distance of one foot from the surface. This will be recognized as an enormous intensity of illumination. On a cloudy day the intensity of illumination at the earth's surface may be as high as 3000 foot-candles and on a "gloomy" day the illumination at the earth's surface may be 1000 foot-candles. When it is considered that mankind works under artificial light with an intensity of only a few foot-candles, the marvels of the visual apparatus are apparent. But it should be noted that the eyes of the human race evolved under natural light. They have been used to great intensities when called upon for their greatest efforts. The human being is wonderfully adaptive, but it could scarcely be hoped that the eyes could readjust themselves in a few generations to the changed conditions of low-intensity artificial lighting. There is no complaint against the range of intensities to which the eye responds, for in range of sensibility it is superior to any man-made device.

For extremely low brightnesses another set of physiological processes come into play. Based purely upon the physiological laws of vision it seems reasonable to conclude that mankind should not work under artificial illumination as low as has been considered necessary owing to the cost in the past. With this principle of vision as a foundation, experiments have been made with greater intensities of illumination in the industries and elsewhere and increased production has been the result. In a test in a factory where an adequate record of production was in effect it was found that an increase in the intensity of illumination from 4 to 12 foot-candles increased the production in various operations. The lowest increase in production was 8 per cent., the highest was 27 per cent., and the average was 15 per cent. The original lighting in this case was better than that of the typical industrial conditions, so that it seems reasonable to expect a greater increase in production when a change is made from the average inadequate lighting of a factory to a well-designed lighting-system giving a high intensity of illumination.

In another test the production under a poor system of lighting by means of bare lamps on drop-cords was compared with that of an excellent system in which well-designed reflectors were used. The intensity of illumination in the latter case was twenty-five times that of the former and the production was increased in various operations from 30 per cent. for the least increase to 100 per cent. for the greatest increase. Inasmuch as the energy consumption in the latter case was increased seven times and the illumination twenty-five times, it is seen that the increase in intensity of illumination was due largely to the use of proper reflectors and to the general layout of the new lighting-system.

In another case a 10 per cent. increase in production was obtained by increasing the intensity of illumination from 3 foot-candles to about 12 foot-candles. This increase of four times in the intensity of illumination involved an increase in consumption of electrical energy of three times the original amount at an increase in cost equal to 1.2 per cent. of the pay-roll. In another test an increase of 10 per cent. in production was obtained at an increase in cost equal to less than 1 per cent. of the payroll. The efficiency of well-designed lighting installations is illustrated in this case, for the illumination intensity was increased six times by doubling the consumption of electrical energy.

Various other tests could be cited, but these would merely emphasize the same results. However, it may be stated that the factory superintendents involved are convinced that adequate and proper artificial lighting is a great factor in increasing production. Mr. W. A. Durgin, who conducted the tests, has stated that the average result of increasing the intensity of illumination and of properly designing the lighting installations in factories will be at least a 15 per cent. increase in production at an increased cost of not more than 5 per cent. of the pay-roll. This is apparently a conservative statement. When it is considered that generally the cost of lighting is only a fraction of 1 per cent. of the cost of products to the consumer, it is seen that the additional cost of obtaining an increase of 15 per cent. in production is inappreciable.

Industrial superintendents are just beginning to see the advantage of adequate artificial lighting, but the low standards of lighting which were inaugurated when artificial light was much more costly than it is to-day persist tenaciously. When high intensities of proper illumination are once tried, they invariably prove successful in the industries. Not only does the worker see all his operations better, but there appears to be an enlivening effect upon individuals under the higher intensities of illumination. Mankind chooses a dimly lighted room in which to rest and to dream. A room intensely lighted by means of well-designed units which are not glaring is comfortable but not conducive to quiet contemplation. It is a place in which to be active. This is perhaps one of the factors which makes for increased production under adequate lighting.

Civilization has just passed the threshold of the age of adequate artificial lighting and only a small percentage of the industries have increased their lighting standards commensurately to the possibilities of the present time. If high-intensity artificial lighting was installed in all the industries and a 15 per cent. increase in production resulted, as tests appear to indicate, the increased production would be equal to that of nearly two million workers. This great increase in output is brought about by lighting at an insignificant increase in cost but without the additional consumption of food or clothing. Besides this increase in production there is the decrease in spoilage. The saving possible in this respect through adequate lighting has been estimated for the industries of this country at $100,000,000. If mankind is to have conveniences and luxuries, efficiency in production must be practised to the utmost and in the foregoing a proved means has been discussed.

There are many other ways in which artificial light may serve in increasing production. Man has found that eight hours of sleep is sufficient to keep him fit for work if he has a sufficient amount of recreation. Before the advent of artificial light the activities of the primitive savage were halted by darkness. This may have been Nature's intention, but civilized man has adapted himself to the changed conditions brought about by efficient and adequate artificial light. There appears to be no fundamental reason for not imposing an artificial day upon plants, animals, chemical processes, etc.; and, in fact, experiments are being prosecuted in these directions.

The hen, when permitted to follow her natural course, rises with the sun and goes to roost at sunset. During the winter months she puts in short days off the roost. It has been shown that an artificial day, made by piecing out daylight by means of artificial light, might keep the hen scratching and feeding longer, with an increased production of eggs as a result. Many experiments of this character have been carried out, and there appears to be a general conclusion that the use of artificial light for this purpose is profitable.

Experiments conducted recently by the agricultural department of a large university indicate that in poultry husbandry, when artificial light is applied to the right kind of stock with correct methods of feeding, the distribution of egg-production throughout the whole year can be radically changed. The supply of eggs may be increased in autumn and winter and decreased in spring and summer. Data on the amount of illumination have not been published, but it is said that the most satisfactory results have been obtained when the artificial illumination is used from sunset until about 9 P. M. throughout the year.

An increase of 30 to 40 per cent. in the number of eggs laid on a poultry-farm in England as the result of installing electric lamps in the hen-houses was reported in 1913. On this farm there were nearly 200 yards of hen-houses containing about 6000 hens, and the runs were lighted on dark mornings and early nights of the year preceding the report. About 300 small lamps varying from 8 to 32 candle-power were used in the houses. It was found that an imitation of sunset was necessary by switching off the 32 candle-power lamps at 6 P. M. and the 16 candle-power lamps at 9:30. This left only the 8 candle-power lamps burning, and in the faint illumination the hens sought the roosting-places. At 10 P. M. the remaining lights were extinguished. It was found that if all the lights were extinguished suddenly the fowls went to sleep on the ground and thus became a prey to parasites. The increase in production of eggs is brought about merely by keeping the fowls awake longer. On the same farm the growth of chicks incubated during the winter months increased by one third through the use of electric light which kept them feeding longer.

Many fishermen will testify that artificial light seems to attract fish, and various reports have been circulated regarding the efficacy of using artificial light for this purpose on a commercial scale. One report which bears the earmarks of authenticity is from Italy, where it is said that electric lights were successfully used as "bait" to augment the supply of fish during the war. The lamps were submerged to a considerable depth and the fish were attracted in such large numbers that the use of artificial light was profitable. The claims made were that the supply of fish was not only increased by night fishing but that a number of fishermen were thereby released for national service during the war. An interesting incident pertaining to fish, but perhaps not an important factor in production, is the use of electric lights in the summer over the reservoirs of a fish hatchery. These lights, which hang low, attract myriads of bugs, many of which fall in the water and furnish natural and inexpensive food for the fish.

Many experiments have been carried out in the forcing of plants by means of artificial light. Some of these were conducted forty years ago, when artificial light was more costly than at the present time. Of course, it is well known that light is essential to plant life and in general it is reasonable to believe that daylight is the most desirable quality of light for plants. In greenhouses the forcing of plants is desirable, owing to the restricted area for cultivation. It has been established that some of the ultra-violet rays which are absorbed or not transmitted by glass are harmful to growing plants. For this reason an arc-lamp designed for forcing purposes should be equipped with a glass globe. F. W. Rane reported in 1894 upon some experiments with electric carbon-filament lamps in greenhouses in which satisfactory results were obtained by using the artificial light several hours each night. Prof. L. H. Bailey also conducted experiments with the arc-lamp and concluded that there were beneficial results if the light was filtered through clear glass. Without considering the details of the experiment, we find some of Rane's conclusions of interest, especially when it is remembered that the carbon-filament lamps used at that time were of very low efficiency compared with the filament lamps at the present time. Some of his conclusions were as follows:

The incandescent electric light has a marked effect upon greenhouse plants.

The light appears to be beneficial to some plants grown for foliage, such as lettuce. The lettuce was earlier, weighed more and stood more erect.

Flowering plants blossomed earlier and continued to bloom longer under the light. The light influences some plants, such as spinach and endive, to quickly run to seed, which is objectionable in forcing these plants for sale.

The stronger the candle-power the more marked the results, other conditions being the same.

Most plants tended toward a taller growth under the light.

It is doubtful whether the incandescent light can be used in the greenhouse from a practical and economic standpoint on other plants than lettuce and perhaps flowering plants; and at present prices (1894) it is a question if it will pay to employ it even for these.

There are many points about the incandescent electric light that appear to make it preferable to the arc light for greenhouse use.

Although we have not yet thoroughly established the economy and practicability of the electric light upon plant growth, still I am convinced that there is a future in it.

These are encouraging conclusions, considering the fact that the cost of light from incandescent lamps at the present time is only a small fraction of its cost at that time.

In an experiment conducted in England in 1913 mercury glass-tube arcs were used in one part of a hothouse and the other part was reserved for a control test. The same kind of seeds were planted in the two parts of the hothouse and all conditions were maintained the same, excepting that a mercury-vapor lamp was operated a few hours in the evening in one of them. Miss Dudgeon, who conducted the test, was enthusiastic over the results obtained. Ordinary vegetable seeds and grains germinated in eight to thirteen days in the hothouse in which the artificial light was used to lengthen the day. In the other, germination took place in from twelve to fifty-seven days. In all cases at least several days were saved in germination and in some cases several weeks. Flowers also increased in foliage, and a 25 per cent. increase in the crop of strawberries was noted. Seedlings produced under the forcing by artificial light needed virtually no hardening before being planted in the open. Professor Priestley of Bristol University said of this work:

The light seems to have been extraordinarily efficacious, producing accelerated germination, increased growth, greater depth of color, and more important still, no signs of lanky, unnatural extension of plant usually associated with forcing. Rather the plants exposed to the radiation seem to have grown if anything more sturdy than the control plants. A structural examination of the experimental and control plants carried out by means of the microscope fully confirmed Miss Dudgeon's statements both as to depth of color and greater sturdiness of the treated plants.

Unfortunately there is much confusion amid the results of experiments pertaining to the effects of different rays, including ultra-violet, visible and infra-red, upon plant growth. If this aspect was thoroughly established, investigations could be outlined to greater advantage and efficient light-sources could be chosen with certainty. There is the discouraging feature that the average intensity of daylight illumination from sunrise to sunset in the summer-time is several thousand foot-candles. The cost of obtaining this great intensity by means of artificial light would be prohibitive. However, the daylight illumination in a greenhouse in winter is very much less than the intensity outdoors in summer. Indeed, this intensity perhaps averages only a few hundred foot-candles in winter. There is encouragement in this fact and there is hope that a little light is relatively much more effective than a great amount. Expressed in another manner, it is possible that a little light is much more effective than no light at all. Experiments with artificial light indicate very generally an increased growth.

Recently Hayden and Steinmetz experimented with a plot of ground 5 feet by 9 feet, over which were hung five 500-watt gas-filled tungsten lamps 3 feet above the ground and 17 inches apart. The lamps were equipped with reflectors and the resulting illumination was 700 foot-candles. This is an extremely high intensity of artificial illumination and is comparable with daylight in greenhouses. The only seeds planted were those of string beans and two beds were carried through to maturity, one lighted by daylight only and the other by daylight and artificial light, the latter being in operation twenty-fours hours per day. The plants under the additional artificial light grew more rapidly than the others, and of the various records kept the gain in time was in all cases about 50 per cent. From the standpoint of profitableness the artificial lighting was not justified. However, there are several points to be brought out before considering this conclusion too seriously. First, it appears unwise to use the artificial light during the day; second, it appears possible that a few hours of artificial light in the evening would suffice for considerable forcing; third, it is possible that a much lower intensity of artificial light might be more effective per lumen than the great intensity used; fourth, it is quite possible that some other efficient light-source may be more effective in forcing the growth of plants. These and many other factors must be carefully determined before judgment can be passed on the efficacy of artificial light in reducing the cost of living in this direction. Certainly, artificial light has been shown to increase the growth of plants and it appears probable that future generations at least will find it profitable to use the efficient light-producers of the coming ages in this manner.

Many other instances could be cited in which artificial light is very closely associated with the cost of living. Overseas shipment of fruit from the Canadian Northwest is responsible for a decided innovation in fruit-picking. In searching for a cause of rotting during shipment it was finally concluded that the temperature at the time of picking was the controlling factor. As a consequence, daytime was considered undesirable for picking and an electric company supplied electric lighting for the orchards in order that the picking might be done during the cool of night. This change is said to have remedied the situation. Cases of threshing and other agricultural operations being carried on at night are becoming more numerous. These are just the beginnings of artificial light in a new field or in a new relation to civilization. Its economic value has been demonstrated in the ordinary fields of lighting and these new applications are merely the initial skirmishes which precede the conquest of new territory. The modern illuminants have been developed so recently that the new possibilities have not yet been established. However, artificial light is already a factor on the side of the people in the struggle against the increasing cost of living, and its future in this direction is still more promising.

XIX

ARTIFICIAL LIGHT AND CHEMISTRY

Some one in an early century was the first to notice that the sun's rays tanned the skin, and this unknown individual made the initial discovery in what is now an extensive branch of science known as photo-chemistry. The fading of dyes, the bleaching of textiles, the darkening of silver salts, the synthesis and decomposition of compounds are common examples of chemical reactions induced by light. There are thousands of other examples of the chemical effects of light some of which have been utilized by mankind. Others await the development of more efficient light-sources emitting greater quantities of active rays, and many still remain interesting scientific facts without any apparent practical applications at the present time. Visible and ultra-violet rays are the radiations almost entirely responsible for photochemical reactions, but the most active of these are the blue, violet, and ultra-violet rays. These are often designated chemical or actinic rays in order to distinguish the group as a whole from other groups such as ultra-violet, visible, and infra-red. Light is a unique agent in chemical reactions because it is not a material substance. It neither contaminates nor leaves a residue. Although much information pertaining to photochemistry has been available for years, the absence of powerful light-sources emitting so-called chemical rays in large quantities inhibited the practical development of the science of photochemistry. Even to-day, with vast applications of light in this manner, mankind is only beginning to utilize its chemical powers.

In a portrait studio

ARTIFICIAL LIGHT IN PHOTOGRAPHY]

City waterworks

STERILIZING WATER WITH RADIANT ENERGY FROM QUARTZ MERCURY-ARCS]

Although it appears that the chemical action of light was known to the ancients, the earliest photochemical investigations which could be considered scientific and systematic were those of K. W. Scheele in 1777 on silver salts. An extract from his own account is as follows:

I precipitated a solution of silver by sal-ammoniac; then I edulcorated (washed) it and dried the precipitate and exposed it to the beams of the sun for two weeks; after which I stirred the powder and repeated the same several times. Hereupon I poured some caustic spirit of sal-ammoniac (strong ammonia) on this, in all appearance, black powder, and set it by for digestion. This menstruum (solvent) dissolved a quantity of luna cornua (horn silver), though some black powder remained undissolved. The powder having been washed was, for the greater part, dissolved by a pure acid of nitre (nitric acid), which, by the operation, acquired volatility. This solution I precipitated again by means of sal-ammoniac into horn silver. Hence it follows that the blackness which the luna cornua acquires from the sun's light, and likewise the solution of silver poured on chalk, is silver by reduction. I mixed so much of distilled water with the well-washed horn silver as would just cover this powder. The half of this mixture I poured into a white crystal phial, exposed it to the beams of the sun, and shook it several times each day; the other half I set in a dark place. After having exposed the one mixture during the space of two weeks, I filtrated the water standing over the horn silver, grown already black; I let some of this water fall by drops in a solution of silver, which was immediately precipitated into horn silver.

This extract shows that Scheele dealt with the reducing action of light. He found that silver chloride was decomposed by light and that there was a liberation of chlorine. However, it was learned later that dried silver chloride sealed in a tube from which the air was exhausted is not discolored by light and that substances must be present to absorb the chlorine. Scheele's work aroused much interest in photochemical effects and many investigations followed. In many of these the superiority of blue, violet, and ultra-violet rays was demonstrated. In 1802 the first photograph was made by Wedgwood, who copied paintings upon glass and made profiles by casting shadows upon a sensitive chemical compound. However, he was not able to fix the image. Much study and experimentation were expended upon photochemical effects, especially with silver compounds, before Niepce developed a method of producing pictures which were subsequently unaffected by light. Later Daguerre became associated with Niepce and the famous daguerreotype was the result. Apparently the latter was chiefly responsible for the development of this first commercial process, the products of which are still to be found in the family album. A century has elapsed since this earliest period of commercial photography, and during each year progress has been made, until at the present time photography is thoroughly woven into the activities of civilized mankind.

In those earliest years a person was obliged to sit motionless in the sun for minutes in order to have his picture taken. The development of a century is exemplified in the "snapshot" of the present time. Photographic exposures outdoors at present are commonly one thousandth of a second, and indoors under modern artificial light miles of "moving-picture" film are made daily in which the individual exposures are very small fractions of a second. Artificial light is playing a great part in this branch of photochemistry, and the development of artificial light for the various photographic needs is best emphasized by reminding the reader that the sources must be generally comparable with the sun in actinic or chemical power. The intensity of illumination due to sunlight on a clear day when the sun is near the zenith is commonly 10,000 foot-candles on a surface perpendicular to the direct rays. This is equivalent to the illumination due to a source 90,000 candle-power at a distance of three feet. The sun delivers about 200,000,000,000 horse-power to the earth continuously, which is estimated to be about one million times the amount of power generated artificially on the earth. Of this inconceivable quantity of energy a small part is absorbed by vegetation, some is reflected and radiated back into space, and the balance heats the earth. To store some of this energy so that it may be utilized at will in any desired form is one of the dreams of science. However, artificial light-sources are depended upon at present in many photographic and other chemical processes.

Although two illuminants may be of the same luminous intensity, they may differ widely in actinic value. It is impossible to rate the different illuminants in a general manner as to actinic value because the various photochemical reactions are not affected to the same extent by rays of a given wave-length. Nearly all human eyes see visible rays in approximately the same manner, but the multitude of chemical reactions show a wide variation in sensitivity to the various rays. For example, one photographic emulsion may be sensitive only to ultra-violet, violet, and blue rays and another to all these rays and also to the green, yellow, and red. Therefore, one illuminant may be superior to another for one photochemical reaction, while the reverse may be true in the case of another reaction. In general, it may be said that the arc-lamps including the mercury-arcs provide the most active illuminants for photochemical processes; however, a large number of electric incandescent filament lamps are used in photographic work.

The photo-engraver has been independent of sunlight since the practical development of his art. In fact, the printer could not depend upon sunlight for making the engravings which are used to illustrate the magazines and newspapers. The newspaper photographer may make a "flashlight" exposure, develop his negative, and make a print from it under artificial light. He may turn this over to the photo-engraver who carries out his work by means of powerful arc-lamps and in an hour or two after the original exposure was made the newspaper containing the illustration is being sold on the streets.

The moving-picture studio is independent of daylight in indoor settings and there is a tendency toward the exclusive use of artificial light. In this field mercury-vapor lamps, arc-lamps, and tungsten photographic lamps are used.

Similarly, in the portrait studio there is a tendency for the photographer to leave the skylighted upper floors and to utilize artificial light. In this field the tungsten photographic lamp is gaining in popularity, owing to its simplicity and to other advantages. Artificial light in general is more satisfactory than natural light for many kinds of photographic work because through the ease of controlling it a greater variety of more artistic effects may be obtained. In ordinary photographic printing tungsten lamps are widely used, but in blue-printing the white flame-arc and the mercury-vapor lamp are generally employed. Not many years ago the blue-printer waited for the sun to appear in order to make his prints, but to-day large machines operate continuously under the light of powerful artificial sources. How many realize that the blue-print is almost universally at the foundation of everything at the present time? Not only are products made from blue-prints but the machinery which makes the products is built from blue-prints. Even the building which houses the machinery is first constructed from blue-prints. They form an endless chain in the activities of present civilization.

Artificial light has been a great factor in the practical development of photography and it is looked upon for aid in many other directions. Although there is a multitude of reactions in photographic processes which are brought about by exposure to light, these represent relatively few of the photochemical reactions. In general, it may be stated that light is capable of causing nearly every type of reaction. The chemical compounds which are photo-sensitive are very numerous. Many of the compounds of silver, gold, platinum, mercury, iron, copper, manganese, lead, nickel, and tin are photo-sensitive and these have been widely investigated. Light and oxygen cause many oxidation reactions and, on the other hand, light reduces many compounds such as silver salts, even to the extent of liberating the metal. Oxygen is converted partially into ozone under the influence of certain rays and there are many examples of polymerization caused by light.

Various allotropic changes of the elements are due to the influence of light; for example, a sulphur soluble in carbon disulphide is converted into sulphur which is insoluble, and the rate of change of yellow phosphorus into the red variety is greatly accelerated by light. Hydrogen and chlorine combine under the action of light with explosive rapidity to form hydrochloric acid and there are many other examples of the synthesizing action of light. Carbon monoxide and chlorine combine to form phosgene and the combination of chlorine, bromine, and iodine, with organic compounds, is much hastened by exposing the mixture to light. In a similar manner many decompositions are due to light; for example, hydrogen peroxide is decomposed into water and oxygen. This suggests the reason for the use of brown bottles as containers for many chemical compounds. Such glass does not transmit appreciably the so-called actinic or chemical rays.

There is a large number of reactions due to light in organic chemistry and one of fundamental importance to mankind is the effect of light on the chlorophyll, the green coloring matter in vegetation. No permanent change takes place in the chlorophyll, but by the action of light it enables the plant to absorb oxygen, carbon dioxide, and water and to use these to build up the complex organic substances which are found in plants. Radiant energy or light is absorbed and converted into chemical energy. This use of radiant energy occurs only in those parts of the plant in which chlorophyll is present, that is, in the leaves and stems. These parts absorb the radiant energy and take carbon dioxide from the air through breathing openings. They convert the radiant energy into chemical energy and use this energy in decomposing the carbon dioxide. The oxygen is exhausted and the carbon enters into the structure of the plant. The energy of plant life thus comes from radiant energy and with this aid the simple compounds, such as the carbon dioxide of the air and the phosphates and nitrates of the soil, are built into complex structures. Thus plants are constructive and synthetic in operation. It is interesting to note that the animal organism converts complex compounds into mechanical and heat energy. The animal organism depends upon the synthetic work of plants, consuming as food the complex structures built by them under the action of light. For example, plants inhale carbon dioxide, liberate the oxygen, and store the carbon in complex compounds, while the animal uses oxygen to burn up the complex compounds derived from plants and exhales carbon dioxide. It is a beautiful cycle, which shows that ultimately all life on earth depends upon light and other radiant energy associated with it. Contrary to most photochemical reactions, it appears that plant life utilize yellow, red, and infra-red energy more than the blue, violet, and ultra-violet.

In general, great intensities of blue light and of the closely associated rays are necessary for most photochemical reactions with which man is industrially interested. It has been found that the white flame-arc excels other artificial light-sources in hastening the chlorination of natural gas in the production of chloroform. One advantage of the radiation from this light-source is that it does not extend far into the ultra-violet, for the ultra-violet rays of short wave-lengths decompose some compounds. In other words, it is necessary to choose radiation which is effective but which does not have rays associated with it that destroy the desired products of the reaction. By the use of a shunt across the arc the light can be gradually varied over a considerable range of intensity. Another advantage of the flame-arc in photochemistry is the ease with which the quality or spectral character of the radiant energy may be altered by varying the chemical salts used in the carbons. For example, strontium fluoride is used in the red flame-arc whose radiant energy is rich in red and yellow. Calcium fluoride is used in the carbons of the yellow flame-arc which emits excessive red and green rays causing by visual synthesis the yellow color. The radiant energy emitted by the snow-white flame-arc is a close approximation to average daylight both as to visible and to ultra-violet rays. Its carbons contain rare-earths. The uses of the flame-arcs are continually being extended because they are of high intensity and efficiency and they afford a variety of color or spectral quality. A million white flame-carbons are being used annually in this country for various photochemical processes.

Of the hundreds of dyes and pigments available many are not permanent and until recent years sunlight was depended upon for testing the permanency of coloring materials. As a consequence such tests could not be carried out very systematically until a powerful artificial source of light resembling daylight was available. It appears that the white flame-arc is quite satisfactory in this field, for tests indicate that the chemical effect of this arc in causing dye-fading is four or five times as great as that of the best June sunlight if the materials are placed within ten inches of a 28-ampere arc. It has been computed that in several days of continuous operation of this arc the same fading results can be obtained as in a year's exposure to daylight in the northern part of this country. Inasmuch as the fastness of colors in daylight is usually of interest, the artificial illuminant used for color-fading should be spectrally similar to daylight. Apparently the white flame-arc fulfils this requirement as well as being a powerful source.

Lithopone, a white pigment consisting of zinc sulphide and barium sulphate, sometimes exhibits the peculiar property of darkening on exposure to sunlight. This property is due to an impurity and apparently cannot be predicted by chemical analysis. During the cloudy days and winter months when powerful sunlight is unavailable, the manufacturer is in doubt as to the quality of his product and he needs an artificial light-source for testing it. In such a case the white flame-arc is serving satisfactorily, but it is not difficult to obtain effects with other light-sources in a short time if an image of the light-source is focused upon the material by means of a lens. In fact, a darkening of lithopone may be obtained in a minute by focusing upon it the image of a quartz mercury-arc by means of a quartz lens. In special cases of this sort the use of a focused image is far superior to the ordinary illumination from the light-source, but, of course, this is impracticable when testing a large number of samples simultaneously. Incidentally, lithopone which turns gray or nearly black in the sunlight regains its whiteness during the night.

An amusing incident is told of a young man who painted his boat one night with a white paint in which lithopone was the pigment. On returning home the next afternoon after the boat had been exposed to sunlight all day, he was astonished to see that it was black. Being very much perturbed, he telephoned to the paint store, but the proprietor escaped a scathing lecture by having closed his shop at the usual hour. The young man telephoned in the morning and told the proprietor what had happened, but on being asked to make certain of the facts he went to the window and looked at his boat and behold! it was white. It had regained whiteness during the night but would turn black again during the day. Although pigments and dyes are not generally as peculiar as lithopone, much uncertainty is eliminated by systematic tests under constant, continuous, and controllable artificial light.

The sources of so-called chemical rays are numerous for laboratory work, but there is a need for highly efficient powerful producers of this kind of energy. In general the flame-arcs perhaps are foremost sources at the present time, with other kinds of carbon arcs and the quartz mercury-arc ranking next. One advantage of the mercury-arc is its constancy. Furthermore, for work with a single wave-length it is easy to isolate one of the spectral lines. The regular glass-tube mercury-arc is an efficient producer of the actinic rays and as a consequence has been extensively used in photographic work and in other photochemical processes. An excellent source for experimental work can be made easily by producing an arc between two small iron rods. The electric spark has served in much experimental work, but the total radiant energy from it is small. By varying the metals used for electrodes a considerable variety in the radiant energy is possible. This is also true of the electric arcs, and the flame-arcs may be varied widely by using different chemical compounds in the carbons.

There are other effects of light which have found applications but not in chemical reactions. For example, selenium changes its electrical resistance under the influence of light and many applications of this phenomenon have been made. Another group of light-effects forms a branch of science known as photo-electricity. If a spark-gap is illuminated by ultra-violet rays, the resistance of the gap is diminished. If an insulated zinc plate is illuminated by ultra-violet or violet rays, it will gradually become positively charged. These effects are due to the emission of electrons from the metal. Violet and ultra-violet rays will cause a colorless glass containing manganese to assume a pinkish color. The latter is the color which manganese imparts to glass and under the influence of these rays the color is augmented. Certain ultra-violet rays also ionize the air and cause the formation of ozone. This can be detected near a quartz mercury-arc, for example, by the characteristic odor.

The foregoing are only a few of the multitude of photochemical reactions and other effects of radiant energy. The development of this field awaits to some extent the production of so-called actinic rays more efficiently and in greater quantities, but there are now many practical applications of artificial light for these purposes. In the extensive fields of photography various artificial light-sources have served for many years and they are constantly finding more applications. Artificial light is now used to a considerable extent in the industries in connection with chemical processes, but little information is available, owing to the secrecy attending these new developments in industrial processes. However, this brief chapter has been introduced in order to indicate another field of activity in which artificial light is serving. It is agreed by scientists that photochemistry has a promising future. Mankind harnesses nature's forces and produces light and this light is put to work to exert its influence for the further benefit of mankind. Science has been at work systematically for only a century, but the accomplishments have been so wonderful that the imagination dares not attempt to prophesy the achievements of the next century.

LIGHT AND HEALTH

The human being evolved without clothing and the body was bathed with light throughout the day, but civilization has gone to the other extreme of covering the body with clothing which keeps most of it in darkness. Inasmuch as light and the invisible radiant energy which is associated with it are known to be very influential agencies in a multitude of ways, the question arises: Has this shielding of the body had any marked influence upon the human organism? Although there is a vast literature upon the subject of light-therapy, the question remains unanswered, owing to the conflicting results and the absence of standardization of experimental details. In fact, most investigations are subject to the criticism that the data are inadequate. Throughout many centuries light has been credited with various influences upon physiological processes and upon the mind. But most of the early applications had no foundation of scientific facts. Unfortunately, many of the claims pertaining to the physiological and psychological effects of light at the present time are conflicting and they do not rest upon an established scientific foundation. Furthermore some of them are at variance with the possibilities and an unprejudiced observer must conclude that much systematic work must be done before order may arise from the present chaos. This does not mean that many of the effects are not real, for radiant energy is known to cause certain effects, and viewing the subject broadly it appears that light is already serving humanity in this field and that its future is promising.

The present lack of definite data pertaining to the effects of radiation is due to the failure of most investigators to determine accurately the quantities and wave-lengths of the rays involved. For example, it is easy to err by attributing an effect to visible rays when the effect may be caused by accompanying invisible rays. Furthermore, it may be possible that certain rays counteract or aid the effective rays without being effective alone. In other words, the physical measurements have been neglected notwithstanding the fact that they are generally more easily made than the determinations of curative effects or of germicidal action. Radiant energy of all kinds and wave-lengths has played a part in therapeutics, so it is of interest to indicate them according to wave-length or frequency. These groups vary in range of wave-length, but the actual intervals are not particularly of interest here. Beginning with radiant energy of highest frequencies of vibration and shortest wave-lengths, the following groups and subgroups are given in their order of increasing wave-length:

Roentgen or X-rays, which pass readily through many substances opaque to ordinary light-rays.

Ultra-violet rays, which are divided empirically into three groups, designated as "extreme," "middle," and "near" in accordance with their location in respect to the visible region.

Visible rays producing various sensations of color, such as violet, blue, green, yellow, orange, and red.

Infra-red or the invisible rays bordering on the red rays.

An unknown, unmeasured, or unfilled region between the infra-red and the "electric" waves.

Electric waves, which include a class of electromagnetic radiant energy of long wave-length. Of these the Herzian waves are of the shortest wave-length and these are followed by "wireless" waves. Electric waves of still greater wave-length are due to the slower oscillations in certain electric circuits caused by lightning discharges, etc.

The Roentgen rays were discovered by Roentgen in 1896 and they have been studied and applied very widely ever since. Their great use has been in X-ray photography, but they are also being used in therapeutics. The extreme ultra-violet rays are not available in sunlight and are available only near a source rich in ultra-violet rays, such as the arc-lamps. They are absorbed by air, so that they are studied in a vacuum. These are the rays which convert oxygen into ozone because the former strongly absorbs them. The middle ultra-violet rays are not found in sunlight, because they are absorbed by the atmosphere. They are also absorbed by ordinary glass but are freely transmitted by quartz. The nearer ultra-violet rays are found in sunlight and in most artificial illuminants and are transmitted by ordinary glass. Next to this region is the visible spectrum with the various colors, from violet to red, induced by radiant energy of increasing wave-length. The infra-red rays are sometimes called heat-rays, but all radiant energy may be converted into heat. Various substances transmit and absorb these rays in general quite differently from the visible rays. Water is opaque to most of the infra-red rays. Next there is a region of wave-lengths or frequencies for which no radiant energy has been found. The so-called electric waves vary in wave-length over a great range and they include those employed in wireless telegraphy. All these radiations are of the same general character, consisting of electromagnetic energy, but differing in wave-length or frequency of vibration and also in their effects. In effect they may overlap in many cases and the whole is a chaos if the physical details of quantity and wave-length are not specified in experimental work.

In a haberdashery

JUDGING COLOR UNDER ARTIFICIAL DAYLIGHT]

In an art gallery

ARTIFICIAL DAYLIGHT]

It has been conclusively shown that radiant energy kills bacteria. The early experiments were made with sunlight and the destruction of micro-organisms is generally attributed to the so-called chemical rays, namely, the blue, violet, and ultra-violet rays. It appears in general that the middle ultra-violet rays are the most powerful destroyers. It is certainly established that sunlight sterilizes water, for example, and the quartz mercury-lamp is in daily use for this purpose on a practicable scale. However, there still appears to be a difference of opinion as to the destructive effect of radiant energy upon bacteria in living tissue. It has been shown that the middle ultra-violet rays destroy animal tissue and, for example, cause eye-cataracts. It appears possible from some experiments that ultra-violet rays destroy bacteria in water and on culture plates more effectively in the absence of visible rays than when these attend the ultra-violet rays as in the case of sunlight. This is one of the reasons for the use of blue glass in light-therapy, which isolates the blue, violet, and near ultra-violet rays from the other visible rays. If the infra-red rays are not desired they can be readily eliminated by the use of a water-cell.

There is a vast amount of testimony which proves the bactericidal action of light. Bacteria on the surface of the body are destroyed by ultra-violet rays. Typhus and tubercle bacilli are destroyed equally well by the direct rays from the sun and from the electric arcs. Cultures of diphtheria develop in diffused daylight but are destroyed by direct sunlight. Lower organisms in water are readily killed by the radiation from any light-source emitting ultra-violet rays comparable with those in direct sunlight. From the great amount of data available it appears reasonable to conclude that radiant energy is a powerful bactericidal agency but that the action is due chiefly to ultra-violet rays. It appears also that no bacteria can resist these rays if they are intense enough and are permitted to play upon the bacteria long enough. The destruction of these organisms appears to be a phenomenon of oxidation, for the presence of oxygen appears to be necessary.

The foregoing remarks about the bactericidal action of radiant energy apply only to bacteria in water, in cultures, and on the surface of the body. There is much uncertainty as to the ability of radiant energy to destroy bacteria within living tissue. The active rays cannot penetrate appreciably into such tissue and many authorities are convinced that no direct destruction takes place. In fact, it has been stated that the so-called chemical rays are more destructive to the tissue cells than to bacteria. Finsen, a pioneer in the use of radiant energy in the treatment of disease, effected many wonderful cures and believed that the bacteria were directly destroyed by the ultra-violet rays. However, many have since come to the conclusion that the beneficent action of the rays is due to the irritation which causes an outflow of serum, thus bringing more antibodies in contact with the bacilli, and causing the destruction of the latter. Hot applications appear to work in the same manner.

Primitive beings of the tropics are known to treat open wounds by exposing them to the direct rays of the sun without dressings of any kind. These wounds are usually infected and the sun's rays render them aseptic and they heal readily. Many cases of sores and surgical wounds have been quickly healed by exposure to sunlight. Even red light has been effective, so it has been concluded by some that rays of almost any wave-length, if intense enough, will effect a cure of this character by causing an effusion of serum. It has also been stated that the chemical rays have powers and have been used in this role for many minor operations.

It is said that the Chinese have used red light for centuries in the treatment of smallpox and throughout the Middle Ages this practice was not uncommon. In the oldest book on medicine written in English there is an account of a successful treatment of the son of Edward I for smallpox by means of red light. It is also stated that this treatment was administered throughout the reigns of Elizabeth and of Charles II. Another account states that a few soldiers confined in dark dungeons recovered from smallpox without pitting. Finsen also obtained excellent results in the treatment of this disease by means of red light. However, in this case it appears that the exclusion of the so-called chemical rays favors healing of the postules of smallpox and that the use of red light is therefore a negative application of light-therapy. In other words, the red light plays no part except in furnishing a light which does not inhibit healing.

Although the so-called actinic rays have curative value in certain cases, there are some instances where light-baths are claimed to be harmful. It is said that sun-baths to the naked body are not so popular as they were formerly, except for obesity, gout, rheumatism, and sluggish metabolism, because it is felt that the shorter ultra-violet rays may be harmful. These rays are said to increase the pulse, respiration, temperature, and blood-pressure and may even start hemorrhages and in excessive amounts cause headache, palpitation, insomnia, and anemia. These same authorities condemn sun-baths to the naked body of the tuberculous, claiming that any cures effected are consummated despite the injury done by the energy of short wave-length. There is no doubt that these rays are beneficial in local lesions, but it is believed that the cure is due to the irritation caused by the rays and the consequent bactericidal action of the increased flow of serum, and not to any direct beneficial result on the tissue-cells. Others claim to cure tuberculosis by means of powerful quartz mercury-arcs equipped with a glass which absorbs the ultra-violet rays of shorter wave-lengths. These conclusions by a few authorities are submitted for what they are worth and to show that this phase of light-therapy is also unsettled.

Any one who has been in touch with light-therapy in a scientific role is bound to note that much ignorance is displayed in the use of light in this manner. In fact, it appears safe to state that light-therapy often smacks of quackery. Very mysterious effects are sometimes attributed to radiant energy, which occasionally border upon superstition. Nevertheless, this kind of energy has value, and notwithstanding the chaos which still exists, it is of interest to note some of the equipment which has been used. Some practitioners have great confidence in the electric bath, and elaborate light-baths have been devised. In the earlier years of this kind of treatment the electric arc was conspicuous. Electrodes of carbon, carbon and iron, and iron have been used when intense ultra-violet rays were desired. The quartz mercury-arc of later years supplies this need admirably. Dr. Cleaves, after many years of experience with the electric-arc bath, has stated:

From the administration of an electric-arc bath there is obtained an action upon the skin, the patient experiences a pleasant and slightly prickly sensation. There is produced, even from a short exposure, upon the skin of some patients a slight erythema, while with others there is but little such effect even from long exposures. The face assumes a normal rosy coloring and an appearance of refreshment and repose on emerging from the bath is always observed. From the administration of the electric-arc bath there is also noted the establishment of circulatory changes with a uniform regulation of the heart's action, as evidenced by improved volume and slower pulse rate, the augmentation of the temperature, increased activity of the skin, fuller and slower respiration, gradually increased respiratory capacity, and diminished irritability of the mucous membrane in tubercular, bronchitic, or asthmatic patients. There is also lessened discharge in those patients suffering from catarrhal conditions of the nasal passages. In diseases of the respiratory system, a soothing effect upon the mucous membranes is always experienced, while cough and expectoration are diminished.

The cabinet used by Dr. Cleaves was large enough to contain a cot upon which the patient reclined. An arc-lamp was suspended at each of the two ends of the cabinet and a flood of light was obtained directly and by reflection from the white inside surfaces of the cabinet. By means of mirrors the light from the arcs could be concentrated upon any desired part of the patient.

Finsen, who in 1895 published his observations upon the stimulating action of light, is considered the pioneer in the use of so-called chemical rays in the treatment of disease. He had a circular room about thirty-seven feet in diameter, in which two powerful 100-ampere arc-lamps about six feet from the floor were suspended from the ceiling. Low partitions extended radially from the center, so that a number of patients could be treated simultaneously. The temperature of the room was normal, so that the treatment was essentially by radiant energy and not by heat. The chemical action upon the skin was said to be quite as strong as under sunlight. The exposures varied from ten minutes to an hour.

Light-baths containing incandescent filament lamps are also used. In some cases the lamp, sometimes having a blue bulb, is merely contained as a reflector and the light is applied locally as desired. Light-cabinets are also used, but in these there is considerable effect due to heat. The ultra-violet rays emitted by the small electric filament lamps used in these cabinets are of very low intensity and the bactericidal action of the light must be feeble. The glass bulbs do not transmit the extreme ultra-violet rays responsible for the production of ozone, or the middle ultra-violet rays which are effective in destroying animal tissue. The cabinets contain from twenty to one hundred incandescent filament lamps of the ordinary sizes, from 25 to 60 watts. In the days of the carbon filament lamp the 16-candle-power lamp was used. Certainly the heating effect has advantages in some cases over other methods of heating. The light-rays penetrate the tissue and are absorbed and transformed into heat. Other methods involve conduction of heat from the hot air or other hot applications. Of course, it is also contended that the light-rays are directly beneficial.

Light is also concentrated upon the body by means of lenses and mirrors. For this purpose the sun, the arc, the quartz mercury-arc, and the incandescent lamp have been used. Besides these, vacuum-tube discharges and sparks have been utilized as sources for radiant energy and "electrical" treatment. Roentgen rays and radium have also figured in recent years in the treatment of disease.

The quartz mercury-arc has been extensively used in the past decade for the treatment of skin diseases and there appears to be less uncertainty about the efficacy of radiant energy for the treatment of surface diseases than of others. Herod related that the Egyptians treated patients by exposure to direct sunlight and throughout the centuries and among all types of civilization sunlight has been recognized as having certain valuable healing or purifying properties. Finsen in his early experiments cured a case of lupus, a tuberculous skin disease, by means of the visible and near ultra-violet rays in sunlight. He demonstrated that these were the effective rays by using only the radiant energy which passed through a water-cell made by using a convex lens for each end of the cell and filling the intervening space with water. This was really a lens made of glass and water. The glass absorbed the ultra-violet rays of shorter wave-length and the water absorbed the infra-red rays. Thus he was able to concentrate upon the diseased skin radiant energy consisting of visible and near ultra-violet rays.

The encouraging results which Finsen obtained in the treatment of skin diseases led him to become independent of sunlight by equipping a special arc-lamp with quartz lenses. This gave him a powerful source of so-called chemical rays, which could be concentrated wherever desired. However, when science contributed the mercury-vapor arc, developments were immediately begun which aimed to utilize this artificial source of steady powerful ultra-violet rays in light-therapy. As a consequence, there are now available very compact quartz mercury-arcs designed especially for this purpose. Apparently their use has been very effective in curing many skin diseases. Certainly if radiant energy is effective, it has a great advantage over drugs. An authority has stated in regard to skin diseases that,

treatment with the ultra-violet rays, especially in conjunction with the Roentgen rays, radium and mesothorium is that treatment which in most instances holds rank as the first, and in many as the only and often enough the most effective mode of handling the disease.

Sterilization by means of the radiation from the quartz mercury-arc has been practised successfully for several years. Compact apparatus is in use for the sterilization of water for drinking, for surgical purposes, and for swimming-pools, and the claims made by the manufacturers of the apparatus apparently are substantiated. One type of apparatus withstands a pressure of one hundred pounds per square inch and may be connected in series with the water-main. The water supplied to the sterilizer should be clear and free of suspended matter, in order that the radiant energy may be effective. Such apparatus is capable of sterilizing any quantity of water up to a thousand gallons an hour, and the lamp is kept burning only when the water is flowing. It is especially useful in hotels, stores, factories, on ships, and in many industries where sterile water is needed.

Water is a vital necessity in every-day life, whether for drinking, cooking, or industrial purposes. It is recognized as a carrier of disease and the purification of water-supply in large cities is an important problem. Chlorination processes are in use which render the treated water disagreeable to the taste and filtration alone is looked upon with suspicion. The use of chemicals requires constant analysis, but it is contended that the bactericidal action of ultra-violet rays is so certain and complete that there is never any doubt as to the sterilization of the water if it is clear, or if it has been properly filtered before treating. The system of sterilization by ultra-violet rays is the natural way, for the sun's rays perform this function in nature. Apparatus for sterilization of water by means of ultra-violet rays is built for public plants in capacities up to ten million gallons per day and these units may be multiplied to meet the needs of the largest cities. Large mechanical filters are used in conjunction with these sterilizers, and thus mankind copies nature's way, for natural supplies of pure water have been filtered through sand and have been exposed to the rays of the sun which free it from germ life.

Some sterilizers of this character are used at the place where a supply of pure water is desired or at a point where water is bottled for use in various parts of a factory, hospital, store, or office building. These were used in some American hospitals during the recent war, where they supplied sterilized water for drinking and for the antiseptic bathing of wounds. In warfare the water supply is exceedingly important. For example, the Japanese in their campaign in Manchuria boiled the water to be used for drinking purposes. The mortality of armies in many previous wars was often much greater from preventable diseases than from bullets, but the Japanese in their war with Russia reversed the mortality statistics. Of a total mortality of 81,000 more than 60,000 died of casualties in battle.

The sterilization of water for swimming-pools is coming into vogue. Heretofore it was the common practice to circulate the water through a filter, in order to remove the impurities imparted to it by the bathers and to return it to the pool. It is insisted by the adherents of sterilization that filtration of this sort is likely to leave harmful bacteria in the water. Sterilizers in which ultra-violet rays are the active rays are now in use for this purpose, being connected beyond the outflow from the filter. The effectiveness of the apparatus has been established by the usual method of counting the bacteria. Near the outlet of the ordinary filter a count revealed many thousand bacteria per cubic inch of water and among these there were bacteria of intestinal origin. Then a sterilizer was installed in which the effective elements were two quartz mercury-lamps which consumed 2.2 amperes each at 220 volts. A count of bacteria in the water leaving the sterilizer showed that these organisms had been reduced to 5 per cent. and finally to a smaller percentage of their original value, and that all those of intestinal origin had been destroyed. In fact, the water which was returned to the pool was better than that which most persons drink. Radiant energy possesses advantages which are unequaled by other bactericidal agents, in that it does not contaminate or change the properties of the water in any way. It does its work of destroying bacteria and leaves the water otherwise unchanged.

These glimpses of the use of the radiant energy as a means of regaining and retaining good health suggest greater possibilities when the facts become thoroughly established and correlated. The sun is of primary importance to mankind, but it serves in so many ways that it is naturally a compromise. It cannot supply just the desired radiant energy for one purpose and at the same time serve for another purpose in the best manner. It is obscured on cloudy days and disappears nightly. These absences are beneficial to some processes, but man in the highly organized activity of present civilization desires radiant energy of various qualities available at any time. In this respect artificial light is superior to the sun and is being improved continually.

XXI

MODIFYING ARTIFICIAL LIGHT

In a single century science has converted the dimly lighted nights with their feeble flickering flames into artificial daytime. In this brief span of years the production of light has advanced far from the primitive flames in use at the beginning of the nineteenth century, but, as has been noted in another chapter, great improvements in light-production are still possible. Nevertheless, the wonderful developments in the last four decades, which created the arc-lamps, the gas-mantle, the mercury-vapor lamps, and the series of electric incandescent-filament lamps, have contributed much to the efficiency, safety, health, and happiness of mankind.

A hundred years ago civilization was more easily satisfied and an improvement which furnished more light at the same cost was all that could be desired. To-day light alone is not sufficient. Certain kinds of radiant energy are required for photography and other photochemical processes and a vast array of colored light is demanded for displays and for effects upon the stage. Man now desires lights of various colors for their expressive effects. He is no longer satisfied with mere light in adequate quantities; he desires certain qualities. Furthermore, he no longer finds it sufficient to be independent of daylight merely in quantity of light. In fact, he has demanded artificial daylight.

Doubtless the future will see the production of efficient light of many qualities or colors, but to-day many of the demands must be met by modifying the artificial illuminants which are available. Vision is accomplished entirely by the distinction of brightness and color. An image of any scene or any object is focused upon the retina as a miniature map in light, shade, and color. Although the distinction of brightness is a more important function in vision than the ability to distinguish colors, color-vision is far more important in daily life than is ordinarily appreciated. One may go through life color-blind without suffering any great inconvenience, but the divine gift of color-vision casts a magical drapery over all creation. Relatively few are conscious of the wonderful drapery of color, except for occasional moments when the display is unusual. Nevertheless a study of vision in nearly all crafts reveals the fact that the distinction of colors plays an important part.

In the purchase of food and wearing-apparel, in the decoration of homes and throughout the arts and industries, mankind depends a great deal upon the appearance of colors. He depends upon daylight in this respect and unconsciously often, when daylight fails, ceases work which depends upon the accurate distinction of colors. His color-vision evolved under daylight; arts and industries developed under daylight; and all his associations of color are based primarily upon daylight. For these reasons, adequate artificial illumination does not make mankind independent of daylight in the practice of arts and crafts and in many minor activities. In quality or spectral character, the unmodified illuminants used for general lighting purposes differ from daylight and therefore do not fully replace it. Noon sunlight contains all the spectral colors in approximately the same proportions, but this is not true of these artificial illuminants. For these reasons there is a demand for artificial daylight.

The "vacuum" tube affords a possibility of an extensive variety of illuminants differing widely in spectral character or color. Every gas when excited to luminescence by an electric discharge in the "vacuum" tube (containing the gas at a low pressure) emits light of a characteristic quality or color. By varying the gas a variety of illuminants can be obtained, but this means of light-production has not been developed to a sufficiently practicable state to be satisfactory for general lighting. Nitrogen yields a pinkish light and the nitrogen tube as developed by Dr. Moore was installed to some extent a few years ago. Neon yields an orange light and has been used in a few cases for displays. Carbon dioxide furnishes a white light similar to daylight and small tubes containing this gas are in use to-day where accurate discrimination of color is essential.

The flame-arcs afford a means of obtaining a variety of illuminants differing in spectral character or color. By impregnating the carbons with various chemical compounds the color of the flame can be widely altered. The white flame-arc obtained by the use of rare-earth compounds in the carbons provides an illuminant closely approximating average daylight. By using various substances besides carbon for the electrodes, illuminants differing in spectral character can be obtained. These are usually rich in ultra-violet rays and therefore have their best applications in processes demanding this kind of radiant energy. The arc-lamp is limited in its application by its unsteadiness, its bulkiness, and the impracticability of subdividing it into light-sources of a great range of luminous intensities.

The most extensive applications of artificial daylight have been made by means of the electric incandescent filament lamp, equipped with a colored glass which alters the light to the same quality as daylight. The light from the electric filament lamp is richer in yellow, orange, and red rays than daylight, and by knowing the spectral character of the two illuminants and the spectral characteristics of colored glasses in which various chemicals have been incorporated, it is possible to develop a colored glass which will filter out of the excess of yellow, orange, and red rays so that the transmitted light is of the same spectral character as daylight. Thousands of such artificial daylight units are now in use in the industries, in stores, in laboratories, in dye-works, in print-shops, and in many other places. Currency and Liberty Bonds have been made under artificial daylight and such units are in use in banks for the detection of counterfeit currency. The diamond expert detects the color of jewels and the microscopist is certain of the colors of his stains under artificial daylight. The dyer mixes his dyes for the coloring of tons of valuable silk and the artist paints under this artificial light. These are only a few of a vast number of applications of artificial daylight, but they illustrate that mankind is independent of natural light in another respect.

There are various kinds of daylight, two of which are fairly constant in spectral character. These are noon sunlight and north skylight. The former may be said to be white light and its spectrum indicates the presence of visible radiant energy of all wave-lengths in approximately equal proportions. North skylight contains an excess of violet, blue, and blue-green rays and as a consequence is a bluish white. Noon sunlight on a clear day is fairly constant in spectral character, but north skylight varies somewhat depending upon the absence or presence of clouds and upon the character of the clouds. If large areas of sunlit clouds are present, the light is largely reflected sunlight. If the sky is overcast, the north skylight is a result of a mixture of sunlight and blue skylight filtered through the clouds and is slightly bluish. If the sky is clear, the light varies from light blue to deep blue.

The daylight which enters buildings is often considerably altered in color by reflection from other buildings and from vegetation, and after it enters a room it is sometimes modified by reflection from colored surroundings. It may be commonly noted that the light reflected from green grass through a window to the upper part of a room is very much tinted with green and the light reflected from a yellow brick building is tinted yellow. Besides these alterations, sunlight varies in color from the yellow or red of dawn through white at noon to orange or red at sunset. Throughout the day the amount of light from the sky does not change nearly as much as the amount of sunlight, so there is a continual variation in the proportion of direct sunlight and skylight reaching the earth. This is further varied by the changing position of the sun. For example, at a north window in which the direct sunlight may not enter throughout the day, the amount of sunlight which enters by reflection from adjacent buildings and other objects may vary greatly. Thus it is seen that daylight not only varies in quantity but also in quality, and an artificial daylight, which is based upon an extensive analysis, has the advantage of being constant in quantity and quality as well as correct in quality. Modern artificial-daylight units which have been scientifically developed not only make mankind independent of daylight in the discrimination of colors but they are superior to daylight.

Although there are many expert colorists who require an accurate artificial daylight, there are vast fields of lighting where a less accurate daylight quality is necessary. The average eyes are not sufficiently skilled for the finest discrimination of colors and therefore the Mazda "daylight" lamp supplies the less exacting requirements of color matching. It is a compromise between quality and efficiency of light and serves the purpose so well that millions of these lamps have found applications in stores, offices, and industries. In order to make an accurate artificial north skylight for color-work by means of colored glass, from 75 to 85 per cent. of the light from a tungsten lamp must be filtered out. This absorption in a broad sense increases the efficiency of the light, for the fraction that remains is now satisfactory, whereas the original light is virtually useless for accurate color-discrimination. About one third of the original light is absorbed by the bulb of the tungsten "daylight" lamp, with a resultant light which is an approximation to average daylight.

Old illuminants such as that emitted by the candle and oil-lamp were used for centuries in interiors. All these illuminants were of a warm yellow color. Even the earlier modern illuminants were not very different in color, so it is not surprising that there is a deeply rooted desire for artificial light in the home and in similar interiors of a warm yellow color simulating that of old illuminants. The psychological effect of warmth and cheerfulness due to such illuminants or colors is well established. Artificial light in the home symbolizes independence of nature and protection from the elements and there is a firm desire to counteract the increasing whiteness of modern illuminants by means of shades of a warm tint. The white light is excellent for the kitchen, laundry, and bath-room, and for reading-lamps, but the warm yellow light is best suited for making cozy and cheerful the environment of the interiors in which mankind relaxes. An illuminant of this character can be obtained efficiently by using a properly tinted bulb on tungsten filament lamps. By absorbing about one fourth to one third of the light (depending upon the temperature of the filament) the color of the candle flame may be simulated by means of a tungsten filament lamp. Some persons are still using the carbon-filament lamp despite its low efficiency, because they desire to retain the warmth of tint of the older illuminants. However, light from a tungsten lamp may be filtered to obtain the same quality of light as is emitted by the carbon filament lamp by absorbing from one fifth to one fourth of the light. The luminous efficiency of the tungsten lamp equipped with such a tinted bulb is still about twice as great as that of the carbon-filament lamp. Thus the high efficiency of the modern illuminants is utilized to advantage even though their color is maintained the same as the old illuminants.

All modern illuminants emit radiant energy, which does not affect the ordinary photographic plate. This superfluous visible energy merely contributes toward glare or a superabundance of light in photographic studios. A glass has been developed which transmits virtually all the rays that affect the ordinary photographic plate and greatly reduces the accompanying inactive rays. Such a glass is naturally blue in color, because it must transmit the blue, violet, and near ultra-violet rays. Its density has been so determined for use in bulbs for the high-efficiency tungsten lamps that the resultant light appears approximately the color of skylight without sacrificing an appreciable amount of the value of the radiant energy for ordinary photography. This glass, it is seen, transmits the so-called chemical rays and is useful in other activities where these rays alone are desired. It is used in light-therapy and in some other activities in which the chemical effects of these rays are utilized.

In the photographic dark-room a deep red light is safe for all emulsions excepting the panchromatic, and lamps of this character are standard products. An orange light is safe for many printing papers. Panchromatic plates and films are usually developed in the dark where extreme safety is desired, but a very weak deep red light is not unsafe if used cautiously. However, many photographic emulsions of this character are not very sensitive to green rays, so a green light has been used for this purpose.

A variety of colored lights are in demand for theatrical effects, displays, spectacular lighting, signaling, etc., and there are many superficial colorings available for this purpose. Few of these show any appreciable degree of permanency. Permanent superficial colorings have recently been developed, but these are secret processes unavailable for the market. For this reason colored glass is the only medium generally available where permanency is desired. For permanent lighting effects, signal glasses, colored caps, and sheets of colored glass may be used. Tints may be obtained by means of colored reflectors. Other colored media are dyes in lacquers and in varnishes, colored inks, colored textiles, and colored pigments.

Inasmuch as colored glass enters into the development of permanent devices, it may be of interest to discuss briefly the effects of various metallic compounds which are used in glass. The exact color produced by these compounds, which are often oxides, varies slightly with the composition of the glass and method of manufacture, but this phase is only of technical interest. The coloring substances in glass may be divided into two groups. The first and largest group consists of those in which the coloring matter is in true solution; that is, the coloring is produced in the same manner as the coloring of water in which a chemical salt is dissolved. In the second group the coloring substances are present in a finely divided or colloidal state; that is, the coloring is due to the presence of particles in mechanical suspension. In general, the lighter elements do not tend to produce colored glasses, but the heavier elements in so far as they can be incorporated into glass tend to produce intense colors. Of course, there are exceptions to this general statement.

The alkali metals, such as sodium, potassium, and lithium, do not color glass appreciably, but they have indirect effects upon the colors produced by manganese, nickel, selenium, and some other elements. Gold in sufficient amounts produces a red in glass and in low concentration a beautiful rose. It is present in the colloidal state. In the manufacture of "gold" red glass, the glass when first cooled shows no color, but on reheating the rich ruby color develops. The glass is then cooled slowly. The gold is left in a colloidal state. Copper when added to a glass produces two colors, blue-green and red. The blue-green color, which varies in different kinds of glasses, results when the copper is fully oxidized, and the red by preventing oxidation by the presence of a reducing agent. This red may be developed by reheating as in the case of making gold ruby glass. Selenium produces orange and red colors in glass.

Silver when applied to the surface of glass produces a beautiful yellow color and it has been widely used in this manner. It has little coloring effect in glass, because it is so readily reduced, resulting in a metallic black. Uranium produces a canary yellow in soda and potash-lime glasses, which fluoresce, and these glasses may be used in the detection of ultra-violet rays. The color is topaz in lead glass. Both sulphur and carbon are used in the manufacture of pale yellow glasses. Antimony has a weak effect, but in the presence of much lead it is used for making opaque or translucent yellow glasses. Chromium produces a green color, which is reddish in lead glass, and yellowish in soda, and potash-lime glasses.

Iron imparts a green or bluish green color to glass. It is usually present as an impurity in the ingredients of glass and its color is neutralized by adding some manganese, which produces a purple color complementary to the bluish green. This accounts for the manganese purple which develops from colorless glass exposed to ultra-violet rays. Iron is used in "bottle green" glass. Its color is greenish blue in potash-lime glass, bluish green in soda-lime glass, and yellowish green in lead glass.

Cobalt is widely used in the production of blue glasses. It produces a violet-blue in potash-lime and soda-lime glasses and a blue in lead glasses. It appears blue, but it transmits deep red rays. For this reason when used in conjunction with a deep red glass, a filter for only the deepest red rays is obtained. Nickel produces an amethyst color in potash-lime glass, a reddish brown in soda-lime glass, and a purple in lead glass. Manganese is used largely as a "decolorizing" agent in counteracting the blue-green of iron. It produces an amethyst color in potash-lime glass and reddish violet in soda-lime and lead glasses.

These are the principal coloring ingredients used in the manufacture of colored glass. The staining of glass is done under lower temperatures, so that a greater variety of chemical compounds may be used. The resulting colors of metals and metallic oxides dissolved in glass depend not only upon the nature of the metal used, but also partly upon the stage of oxidation, the composition of the glass and even upon the temperature of the fusion.

In developing a glass filter the effects of the various coloring elements are determined spectrally and the various elements are varied in proper proportions until the glass of desired spectral transmission is obtained. It is seen that the coloring elements are limited and the combination of these is further limited by chemical considerations. In combining various colored glasses or various coloring elements in the same glass the "subtractive" method of color-mixture is utilized. For example, if a green glass is desired, yellowish green chromium glass may be used as a basis. By the addition of some blue-green due to copper, the yellow rays may be further subdued so that the resulting color is green.

The primary colors for this method of color-mixture are the same as those of the painter in mixing pigments--namely, purple, yellow, and blue-green. Various colors may be obtained by superposing or intimately mixing the colors. The resulting transmission (reflection in the case of reflecting media such as pigments) are those colors commonly transmitted by all the components of a mixture. Thus,

Purple and yellow = red Yellow and blue-green = green Blue-green and purple = blue

The colors produced by adding lights are based not on the "subtractive" method but on the actual addition of colors. These primaries are red, green, and blue and it will be noted that they are the complementaries of the "subtractive" primaries. By the use of red, green, and blue lights in various proportions, all colors may be obtained in varying degrees of purity. The chief mixtures of two of the "additive" primaries produce the "subtractive" primaries. Thus,

Red and blue = purple Red and green = yellow Green and blue = blue-green

Although the coloring media which are permanent under the action of light, heat, and moisture are relatively few, by a knowledge of their spectral characteristics and other principles of color the expert is able to produce many permanent colors for lighting effects. The additive and subtractive methods are chiefly involved, but there is another method which is an "averaging" additive one. For example, if a warm tint of yellow is desired and only a dense yellow glass is available, the yellow glass may be cut into small pieces and arranged upon a colorless glass in checker-board fashion. Thus a great deal of uncolored light which is transmitted by the filter is slightly tinted by the yellow light passing through the pieces of yellow glass. If this light is properly mixed by a diffusing glass the effect is satisfactory. These are the principal means of obtaining colored light by means of filters and by mixing colored lights. By using these in conjunction with the array of light-sources available it is possible to meet most of the growing demands. Of course, the ideal solution is to make the colored light directly at the light-source, and doubtless future developments which now appear remote or even impossible will supply such colored illuminants. In the meantime, much is being accomplished with the means available.

XXII

SPECTACULAR LIGHTING

Artificial light is a natural agency for producing spectacular effects. It is readily controlled and altered in color and the brightness which it lends to displays outdoors at night renders them extremely conspicuous against the darkness of the sky. It surpasses other decorative media by the extreme range of values which may be obtained. The decorator and painter are limited by a range of values from black to white pigments, which ordinarily represents an extreme contrast of about one to thirty. The brightnesses due to light may vary from darkness to those of the light-sources themselves. The decorator deals with secondary light--that is, light reflected by more or less diffusely reflecting objects. The lighting expert has at his command not only this secondary light but the primary light of the sources. Lighting effects everywhere attract attention and even the modern merchant testifies that adequate lighting in his store is of advertising value. In all the field of spectacular lighting the superiority of artificial light over natural light is demonstrated.

Light is a universal medium with which to attract attention and to enthrall mankind. The civilizations of all ages have realized this natural power of light. It has played a part in the festivals and triumphal processions from time immemorial and is still the most important feature of many celebrations. In the early festivals fires, candles, and oil-lamps were used and fireworks were invented for the purpose. Even to-day the pyrotechnical displays against the dark depths of the night sky hold mankind spellbound. But these evanescent notes of light have been improved upon by more permanent displays on a huge scale. Thirty years before the first practical installation of gas-lighting an exhibition of "Philosophical Fireworks" produced by the combustion of inflammable gases was given in several cities of England.

It is a long step from the array of flickering gas-flames with which the fronts of the buildings of the Soho works were illuminated a century ago to the wonderful lighting effects a century later at the Panama-Pacific Exposition. Some who saw that original display of gas-jets totaling a few hundred candle-power described it as an "occasion of extraordinary splendour." What would they have said of the modern spectacular lighting at the Exposition where Ryan used in a single effect forty-eight large search-lights aggregating 2,600,000,000 beam candle-power! No other comparison exemplifies more strikingly the progress of artificial lighting in the hundred years which have elapsed since it began to be developed.

The nature of the light-sources in the first half of the nineteenth century did not encourage spectacular or display lighting. In fact, this phase of lighting chiefly developed along with electric lamps. Of course, occasionally some temporary effect was attempted as in the case of illuminating the dome of St. Paul's Cathedral in London in 1872, but continued operation of the display was not entertained. In the case of lighting this dome a large number of ship's lanterns were used, but the result was unsatisfactory. After this unsuccessful attempt at lighting St. Paul's, a suggestion was made of "flooding it with electric light projected from various quarters." Spectacular lighting outdoors really began in earnest in the dawn of the twentieth century.

Although some of the first attempts at spectacular lighting outdoors were made with search-lights, spectacular lighting did not become generally popular until the appearance of incandescent filament lamps of reasonable efficiency and cost. The effects were obtained primarily by the use of small electric filament lamps draped in festoons or installed along the outlines and other principal lines of buildings and monuments. The effect was almost wholly that of light, for the glare from the visible lamps obscured the buildings or other objects. The method is still used because it is simple and the effects may be permanently installed without requiring any attention excepting to replace burned-out lamps. However, the method has limitations from an artistic point of view because the artistic effects of painting, sculpture, and architecture cannot be combined with it very effectively. For example, the details of a monument or of a building cannot be seen distinctly enough to be appreciated. The effect is merely that of outlines or lines and patterns of points of light and is usually glaring.

The next step was to conceal these lamps behind the cornices or other projections or in nooks constructed the purpose. Light now began to mold and to paint the objects. The structures began to be visible; at least the important cornices and other details were no longer mere outlines. The introduction of the drawn-wire tungsten lamp is responsible for an innovation in spectacular lighting of this sort, for now it became possible to make concentrated light-sources so essential to projectors. Furthermore, these lighting units require very little attention after once being located. With the introduction of electric-filament lamps of this character small projectors came into use, and by means of concentrated beams of light whole buildings and monuments could be flooded with light from remote positions. The effects obtained by concealing lamps behind cornices had demonstrated that the lighting of the surfaces was the object to be realized in most cases, and when small projectors not requiring constant attention became available, a great impetus was given to flood-lighting.

When France gave to this country the Bartholdi Statue of Liberty there was no thought of having this emblem visible at night excepting for the torch held the hand of Liberty. This torch was modified at the time of the erection of the statue to accommodate the lamps available, with the result that it was merely a lantern containing a number of electric lamps. At night it was a speck of light more feeble than many surrounding shore lights. The statue had been lighted during festivals with festoons and outlines of lamps, but in 1915, when the freedom of the generous donor of the statue appeared to be at stake, a movement was begun which culminated in a fund for flood-lighting Liberty. The broad foundation of the statue made the lighting comparatively easy by means of banks of incandescent filament search-lights. About 225 of these units were used with a total beam candle-power of about 20,000,000. The original idea of an imitation flame for the torch was restored by building this from pieces of yellow cathedral glass of three densities. About six hundred pieces of glass were used, the upper ones being generally of the lighter tints and the lower ones of the darker tints. A lighthouse lens was placed in this lantern so that an intense beam of light would radiate from it. The flood-lighted Statue of Liberty is now visible by night as well as by day and it has a double significance at night, for light also symbolizes independence.

Just as the Statue of Liberty stands alone in the New York Harbor so does the Woolworth Building reign supreme on lower Manhattan. Liberty proclaims independence from the bondage of man and the Woolworth Tower stands majestically in defiance of the elements as a symbol of man's growing independence of nature. This building with its cream terra-cotta surface and intricate architectural details touched here and there with buff, blue, green, red, and gold, rises 792 feet or sixty stories above the street and typifies the American spirit of conceiving and of executing great undertakings. In it are blended art, utility, and majesty. Viewed by multitudes during the day, it is a valuable advertisement for the name which stands for a national institution. But by day it shares attention with its surroundings. If lighted at night it would stand virtually alone against the dark sky and the investment would not be wholly idle during the evening hours.

Mr. H. H. Magdsick, who designed the lighting for Liberty, planned the lighting for the Woolworth Tower, which rises 407 feet or thirty-one stories above the main building. Five hundred and fifty projectors containing tungsten filament lamps were distributed about the base of the tower and among some of the architectural details. The main architectural features of the mansard roof extending from the fifty-third to the fifty-seventh floor, the observation balcony at the fifty-eighth and the lantern structures at the fifty-ninth and sixtieth floors are covered with gold-leaf. By proper placing of the projectors a glittering effect is obtained from these gold surfaces. The crowning features of the lighting effect are the lanterns in the crest of the spire. Twenty-four 1000-watt tungsten lamps were placed behind crystal diffusing glass, which transmits the light predominantly in a horizontal direction. Thus at long distances, from which the architectural details cannot be distinguished, the brilliant crowning light is visible. An automatic dimmer was devised so that the effect of a huge varying flame was obtained. At close range, owing to the nature of the glass panels, this portion is not much brighter than the remainder of the surfaces. When the artificial lighting is in operation the tower becomes a majestic spire of light and this magnificent Gothic structure projecting defiantly into the depths of darkness is in more than one sense a torch of modern civilization.

Many prominent buildings and monuments have burst forth in a flood of light, and their beauty and symbolism have been appreciated at night by many persons who do not notice them by day. Not only are the beautiful structures of man lighted permanently but many temporary effects are devised. Artificial lighting effects have become a prominent part in outdoor festivals, pageants, and theatricals. Candles have been associated with Christmas trees ever since the latter came into use and naturally artificial light has been a feature in the community Christmas trees which have come into vogue in recent years. The Municipal Christmas Tree in Chicago in 1916 was ninety feet high and was lighted with projectors. Thousands of gems taken from the Tower of Jewels at the San Francisco Exposition added life and sparkle to that of the other decorations.

Luna Park, Coney Island, studded with 60,000 incandescent filament lamps

THE NEW FLOOD LIGHTING CONTRASTED WITH THE OLD OUTLINE LIGHTING]

After the close of the recent war artificial light played a prominent part throughout the country in the joyful festivals. A jeweled arch erected in New York in honor of the returning soldiers rivaled some of the spectacles of the Panama-Pacific Exposition. The arch hung like a gigantic curtain of jewels between two obelisks, which rose to a height of eighty feet and were surmounted by jeweled forms in the shape of sunbursts. Approximately thirty thousand jewels glittered in the beams of batteries of arc-projectors. Many of the signs and devices which played a part in the "Welcome Home" movement were of striking nature and of a character to indicate permanency. The equipment of a large building consisted of more than five thousand 10-watt lamps, the entire building being outlined with stars consisting of eleven lamps each. The "Brighten Up" campaign spread throughout the country. The lighting and installation of signs and special patriotic displays, the flooding of streets and shop-windows with light without stint, produced an inspiring and uplifting effect which did much to restore cheerfulness and optimism. A glowing example was set in Washington, where the flood-lighting of the Capitol, discontinued shortly after our entrance into the war, was resumed.

In Chicago a "Victory Way" was established, with street-lighting posts on both sides of the street equipped with red, white, and blue globes surmounted by a golden goddess of Victory. One hundred and seventy-five projectors were installed along the way on the roofs and in the windows of office buildings. A brilliant, scintillating "Altar of Victory" was erected at the center of the Way. It was composed of two enormous candelabra erected one on each side of a platform ninety feet high. These were studded with jewels and supported a curtain of jewels suspended from the altar. In the center of the curtain was a huge jeweled eagle bearing the Allied flags. This was illuminated by arc-projectors which delivered 200,000,000 beam candle-power. In addition to these there were many smaller projectors. In the top of each candelabra six large red-and-orange lamps were installed in reflectors. These illuminated live steam which issued from the top. Surmounting the whole was a huge luminous fan formed by beams from large arc search-lights. These are only a few of the many lighting effects which welcomed the returning soldiers, but they illustrate how much modern civilization depends upon artificial light for expressing its feelings and emotions. Throughout all these festivals light silently symbolized happiness, freedom, and advancement.

Projectors were used on a large scale in several cases before the advent of the concentrated filament lamp. W. D'A. Ryan, the leader in spectacular lighting, lighted the Niagara Falls in 1907 with batteries of arc-projectors aggregating 1,115,000,000-beam candle-power. In 1908 he used thirty arc-projectors to flood the Singer Tower in New York with light and projected light to the flag on top by means of a search-light thirty inches in diameter. Many flags waved throughout the war in the beams of search-lights, symbolizing a patriotism fully aroused. The search-light beam as it bores through the atmosphere at night is usually faintly bright, owing to the small amount of fog, dust, and smoke in the air. By providing more "substance" in the atmosphere, the beams are made to appear brighter. Following this reasoning, Ryan developed his scintillator consisting of a battery of search-light beams projected upward through clouds of steam which provided an artificial fog. This was first displayed at the Hudson-Fulton celebration with a battery of arc search-lights totaling 1,000,000,000-candle-power.

All these effects despite their magnitude were dwarfed by those at the Panama-Pacific Exposition, and inasmuch as this up to the present time represents the crowning achievement in spectacular lighting, some of the details worked out by Ryan may be of interest. In general, the lighting effects departed from the bizarre outline lighting in which glaring light-sources studded the structures. The radiant grandeur and beauty of flood-lighting from concealed light-sources was the key-note of the lighting. In this manner wonderful effects were obtained, which not only appealed to the eye and to the artistic sensibility but which were free from glare. By means of flood-lighting and relief-lighting from concealed light-sources the third dimension or depth was obtained and the architectural details and colorings were preserved. A great many different kinds of devices and lamps were used to make the night effects superior in grandeur to those of daytime. The Zone or amusement section was lighted with bare lamps in the older manner and the glaring bizarre effects contrasted the spectacular lighting of the past with the illumination of the future.

In another section the visitor was greeted with a gorgeous display of carnival spirit. Beautifully colored heraldic shields on which were written the early history of the Pacific coast were illuminated by groups of luminous arc-lamps on standards varying from twenty-five to fifty-five feet in height. The Tower of Jewels with more than a hundred thousand dangling gems was flood-lighted, and the myriads of minute reflected images of light-sources glittering against the dark sky produced an effect surpassing the dreams of imagination. Shadows and high-lights of striking contrasts or of elusive colors greeted the visitor on every hand. Individual isolated effects of light were to be found here and there. Fire hissed from the mouths of serpents and cast the spell of mobile light over the composite Spanish-Gothic-Oriental setting. A colored beam of a search-light played here and there. Mysterious vapors rising from caldrons were in reality illuminated steam. Symbolic fountain groups did not escape the magic touch of the lighting wizard.

In the Court of the Universe great areas were illuminated by two fountains rising about a hundred feet above the sunken gardens. One of these symbolized the setting sun, the other the rising sun. The shaft and ball at the crest of each fountain were glazed with heavy opal glass imitating travertine marble and in these were installed incandescent lamps of a total candle-power of 500,000. The balustrade seventy feet above the sunken gardens was surmounted by nearly two hundred incandescent filament search-lights. Light was everywhere, either varying in color into a harmonious scene or changing in light and shadow to mold the architecture and sculpture. The enormous glass dome of the Palace of Horticulture was converted into an astronomical sphere by projecting images upon it in such a manner that spots of light revolved; rings and comets which appeared at the horizon passed on their way through the heavens, changing in color and disappearing again at the horizon. All these effects and many more were mirrored in the waters of the lagoons and the whole was a Wonderland indeed.

The scintillator consisted of 48 arc search-lights three feet in diameter totaling 2,600,000,000 beam candle-power. The lighting units were equipped with colored screens and the beams which radiated upward were supplied with an artificial fog by means of steam generated by a modern express locomotive. The latter was so arranged that the wheels could be driven at a speed of sixty miles per hour under brake, thereby emitting great volumes of steam and smoke, which when illuminated with various colors produced a magnificent spectacle. Over three hundred scintillator effects were worked out and this feature of fireless fireworks was widely varied. The aurora borealis and other effects created by this battery of search-lights extended for many miles. The many effects regularly available were augmented on special occasions and it is safe to state that this apparatus built upon a huge scale provided a flexibility of fireless fireworks never attained even with small-scale devices.

The lighting of the exposition can barely be touched upon in a few paragraphs and it would be difficult to describe in words even if space were unlimited. It represented the power of light to beautify and to awe. It showed the feebleness of the decorator's media in comparison with light pulsating with life. It consisted of a great variety of direct, masked, concealed, and projected effects, but these were blended harmoniously with one another and with the decorative and architectural details of the structures. It was a crowning achievement of a century of public lighting which began with Murdock's initial display of a hundred flickering gas-jets. It demonstrated the powers of science in the production of light and of genius and imagination in the utilization of light. It was a silent but pulsating display of grandeur dwarfing into insignificance the aurora borealis in its most resplendent moments.

XXIII

THE EXPRESSIVENESS OF LIGHT

From an esthetic or, more broadly, a psychological point of view no medium rivals light in expressiveness. Not only is light allied with man's most important sense but throughout long ages of associations and uses mankind has bestowed upon it many attributes. In fact, it is possible that light, color, and darkness possess certain fundamentally innate powers; at least, they have acquired expressive and impressive powers through the many associations in mythology, religion, nature, and common usage. Besides these attributes, light possesses a great advantage over the media of decoration in obtaining brightness and color effects. For example, the landscape artist cannot reproduce the range of values or brightnesses in most of nature's scenes, for if black is used to represent a deep shadow, white is not bright enough to represent the value of the sky. In fact, the range of brightnesses represented by the deep shadow and the sky extends far beyond the range represented by black and white pigments. The extreme contrast ordinarily available by means of artist's colors is about thirty to one, but the sky is a thousand times brighter than a shadow, a sunlit cloud is thousands of times brighter than the deep shadows of woods, and the sun is millions of times brighter than the shadows in a landscape.

The range of brightnesses obtainable by means of light extends from darkness or black throughout the range represented by pigments under equal illumination and beyond these through the enormous range obtainable by unequal illumination of surfaces to the brightnesses of the light-sources themselves. In the matter of purity of colors, light surpasses reflecting media, for it is easy to obtain approximately pure hues by means of light and to obtain pure spectral hues by resorting to the spectrum of light. It is impossible to obtain pure hues by means of pigments or of other reflecting media. These advantages of light are very evident on turning to spectacular lighting effects, and even the lighting of interiors illustrates a potentiality in light superior to other media. For example, in a modern interior in which concealed lighting produces brilliantly illuminated areas above a cornice and dark shadows on the under side, the range in values is often much greater than that represented by black and white, and still there remains the possibility of employing the light-sources themselves in extending the scale of brightness. Superposing color upon the whole it is obvious that the combination of "primary" light with reflected light possesses much greater potentiality than the latter alone. This potentiality of light is best realized if lighting is regarded as "painting with light" in a manner analogous to the decorator's painting with pigments, etc.

The expressive possibilities of lighting find extensive applications in relation to painting, sculpture, and architecture. A painting is an expression of light and the sculptor's product finally depends upon lighting for its effectiveness. Lighting is the master painter and sculptor. It may affect the values of a painting to some extent and it is a great influence upon the colors. It molds the model from which the sculptor works and it molds the completed work.

The direction, distribution, and quality of light influence the appearance of all objects and groups of them. Aside from the modeling of ornament, the light and shade effects of relatively large areas in an interior such as walls and ceiling, the contrasts in the brightnesses of alcoves with that of the main interior, and the shadows under cornices, beams, and arches are expressions of light.

The decorator is able to produce a certain mood in a given interior by varying the distribution of values and the choice of colors and the lighting artist is able to do likewise, but the latter is even able to alter the mood produced by the decorator. For example, a large interior flooded with light from concealed sources has the airiness and extensiveness of outdoors. If lighted solely by means of sources concealed in an upper cornice, the ceiling may be bright and the walls may be relatively dark by contrast. Such a lighting effect may produce a feeling of being hemmed in by the walls without a roof. If the room is lighted by means of chandeliers hung low and equipped with shades in such a manner that the lower portions of the walls may be light while the upper portions of the interior may be ill defined, the feeling produced may be that of being hemmed in by crowding darkness. Thus lighting is productive of moods and illusions ranging from the mystery of crowding darkness to the extensiveness of outdoors.

Future lighting of interiors doubtless will provide an adequacy of lighting effects which will meet the respective requirements of various occasions. A decorative scheme in which light and medium grays are employed produces an interior which is very sensitive to lighting effects. To these light-and-shade effects colored light may add its charming effectiveness. Not only are colored lighting effects able to add much to the beauty of the setting but they possess certain other powers. Blue tints produce a "cold" effect and the yellow and orange tints a "warm" effect. For example, a room will appear cooler in the summer when illuminated by means of bluish light and a practical application of this effect is in the theater which must attract audiences in the summer. How tinted illuminants fit the spirit of an occasion or the mood of a room may be fully appreciated only through experiments, but these are so effective that the future of lighting will witness the application of the idea of "painting with light" to its fullest extent. Color is demanded in other fields, and, considering its effectiveness and superiority in lighting, it will certainly be demanded in lighting when its potentiality becomes appreciated and readily utilized.

The expressiveness of light is always evident in a landscape. On a sunny day the mood of a scene varies throughout the day and it grows more enticing and agreeable as the shadows lengthen toward evening. The artist in painting a desert scene employs short harsh shadows if he desires to suggest the excessive heat. These shadows suggest the relentless noonday sun. The overcast sky is universally depressing and it has been found that on a sunny day most persons experience a slight depression when a cloud obscures the sun. Nature's lighting varies from moment to moment, from day to day, and from season to season. It presents the extremes of variation in distributions of light from overcast to sunny days and in the latter cases the shadows are continually shifting with the sun's altitude. They are harshest at noon and gradually fade as they lengthen, until at sunset they disappear. The colors of sunlit surfaces and of shadows vary from sunrise to sunset. These are the fundamental variations in the lighting, but in the various scenes the lighting effects are further modified by clouds and by local conditions or environment. The vast outdoors provides a fruitful field for the study of the expressiveness of light.

Having become convinced of this power of light, the lighting expert may turn to artificial light, which is so easily controlled in direction, distribution, and color, and draw upon its potentiality. Not only is it easy to provide a lighting suitable to the mood or to the function of an interior but it is possible to obtain some variety in effect so that the lighting may always suit the occasion. A study of nature's lighting reveals one great principle, namely, variety. Mankind demands variety in most of his activities. Work is varied and alternated with recreation. Meals are not always the same. Clothing, decorations, and furnishings are relieved of monotony. One of the most potent features of artificial light is the ease with which variety may be obtained. In obtaining relief from the monotony of decorations and furnishings, considerable expense and inconvenience are inevitably encountered. With an adequate supply of outlets, circuits, and controls a wide variety of lighting effects may be obtained with perhaps an insignificant increase in the initial investment. Variety is the spice of lighting as well as of life.

These various principles of lighting are readily exemplified in the lighting of the home, which is discussed in another chapter. The church is even a better example of the expressive possibilities of lighting. The architectural features are generally of a certain period and first of all it is essential to harmonize the lighting effect with that of the architectural and decorative scheme. Obviously, the dark-stained ceiling of a certain type of church would not be flooded with light. The fact that it is made dark by staining precludes such a procedure in lighting. The characteristics of creeds are distinctly different and these are to some extent exemplified by the lines of the architecture of their churches. In the same way the lighting effect may be harmonized with the creed and the spirit of the interior. The lighting may always be dignified, impressive, and congruous. Few churches are properly lighted with a high intensity of illumination; moderate lighting is more appropriate, for it is conducive to the spirit of worship. In some creeds a dominant note is extreme penitence and severity. The architecture may possess harsh outlines, and this severity or extreme solemnity may be expressed in lighting by harsher contrasts, although this does not mean that the lighting must be glaring. On the other hand, in a certain modern creed the dominant note appears to be cheerfulness. The spacious interiors of the churches of this creed are lacking in severe lines and the walls and ceilings are highly reflecting. Adequate illumination by means of diffused light without the production of severe contrasts expresses the creed, modernity, and enlightenment. On the altar of certain churches the expressiveness of light is utilized in the ceremonial uses which vary with the creed. Even the symbolism of color may be appropriately woven into the lighting of the church.

The expressiveness of light and color originated through the contact of primitive man with nature. Sunlight meant warmth and a bountiful vegetation, but darkness restricted his activities and harbored manifold dangers. Many associations thus originated and they were extended through ignorance and superstition. Yellow is naturally emblematical of the sun and it became the symbol of warmth. Brown as the predominant color of the autumn foliage became tinctured with sadness because the decay of the vegetation presaged the death of the year and the cold dreary months of winter. The first signs of green vegetation in the spring were welcomed as an end of winter and a beginning of another bountiful summer; hence green symbolized youth and hope. It became associated with the springtime of life and thus signified inexperience, but as the color of vegetation it also meant life itself and became a symbol of immortality. Blue acquired certain divine attributes because, as the color of the sky, it was associated with the abode of the gods or heaven. Also a blue sky is the acme of serenity and this color acquired certain appropriate attributes.

Associations of this character became woven into mythology and thus became firmly established. Poets have felt these influences of light and color in nature and have given expression to them in words. They also have entwined much of the mythology of past civilizations and these repetitions have helped to establish the expressiveness of light and color. Early ecclesiasts employed these symbolisms in religious ceremonies and dictated the garbs of saints and other religious personages in the paintings which decorated their edifices. Thus there were many influences at work during the early centuries when intellects were particularly susceptible through superstition and lack of knowledge. The result has been an extensive symbolism of light, color, and darkness.

At the present time it is difficult to separate the innate appeal of light, color, and darkness from those attributes which have been acquired through associations. Possibly light and color have no innate powers but merely appear to have because the acquired attributes have been so thoroughly established through usage and common consent. Space does not permit a discussion of this point, but the chief aim is consummated if the existence of an expressiveness and impressiveness of light is established. There are many other symbolisms of color and light which have arisen in various ways but it is far beyond the scope of this book to discuss them.

Psychological investigations reveal many interesting facts pertaining to the influence of light and color upon mankind. When choosing color for color's sake alone, that is, divorced from any associations of usage, mankind prefers the pure colors to the tints and shades. It is interesting to note that this is in accord with the preference exhibited by uncivilized beings in their use of colors for decorating themselves and their surroundings. Civilized mankind chooses tints and shades predominantly to live with, that is, for the decoration of his surroundings. However, civilized man and the savage appear to have the same fundamental preference for pure colors and apparently culture and refinement are responsible for their difference in choice of colors to live with. This is an interesting discovery and it has its applications in lighting, especially in spectacular and stage-lighting.

It appears to be further established that when civilized man chooses color for color's sake alone he not only prefers the pure colors but among these he prefers those near the ends of the spectrum, such as red and blue. Red is favored by women, with blue a close second, but the reverse is true for men. It is also thoroughly established that red, orange, and yellow exert an exciting influence; yellow-green, green, and blue-green, a tranquilizing influence, and blue and violet a subduing influence upon mankind. All these results were obtained with colors divorced from surroundings and actual usage. In the use of light and color the laws of harmony and esthetics must be obeyed, but the sensibility of the lighting artist is a satisfactory guide. Harmonies are of many varieties, but they may be generally grouped into two classes, those of analogy and those of contrast. The former includes colors closely associated in hue and the latter includes complementary colors. No rules in simplified form can be presented for the production of harmonies in light and color. These simplifications are made only by those who have not looked deeply enough into the subject through observation and experiment to see its complexity.

The expressiveness of light finds applications throughout the vast field of lighting, but the stage offers great opportunities which have been barely drawn upon. When one has awakened to the vast possibilities of light, shade, and color as a means of expression it is difficult to suppress a critical attitude toward the crudity of lighting effects on the present stage, the lack of knowledge pertaining to the latent possibilities of light, and the superficial use of this potential medium. The crude realism and the almost total absence of deep insight into the attributes of light and color are the chief defects of stage-lighting to-day. One turns hopefully toward the gallant though small band of stage artists who are striving to realize a harmony of lighting, setting, and drama in the so-called modern theater. Unappreciated by a public which flocks to the melodramatic movie, whose scenarios produced upon the legitimate stage would be jeered by the same public, the modern stage artist is striving to utilize the potentiality of light. But even among these there are impostors who have never achieved anything worth while and have not the perseverance to learn to extract some of the power of light and to apply it effectively. Lighting suffers in the hands of the artist owing to the absence of scientific knowledge and it is misused by the engineer who does not possess an esthetic sensibility. Science and art must be linked in lighting.

The worthy efforts of stage artists in some of the modern theaters lack the support of the producers, who cater to the taste of the public which pays the admission fees. Apparently the modern theater must first pass through a period in which financial support must be obtained from those who are able to give it, just as the symphony orchestra has been supported for the sake of art. Certainly the time is at hand for philanthropy to come to the aid of worthy and capable stage artists who hope to rescue theatrical production from the mire of commercialism.

Those who have not viewed stage-lighting from behind the scenes would often be surprised at the crudity of the equipment, and especially at the superficial intellects which are responsible for some of the realistic effects obtained. But these are the result usually of experiment, not of directed knowledge. Furthermore, little thought is given to the emotional value of light, shade, and color. The flood of light and the spot of light are varied with gaudy color-effects, but how seldom is it possible to distinguish a deep relation between the lighting and the dramatic incidents!

Jeweled portal welcoming returned soldiers

ARTIFICIAL LIGHT HONORING THOSE WHO FELL AND THOSE WHO RETURNED]

In much of the foregoing discussion the present predominating theatrical productions are not considered, for the lighting effects are good enough for them. Many ingenious tricks and devices are resorted to in these productions, and as a whole lighting is serving effectively enough. But in considering the expressiveness of light the deeper play is the medium necessary for utilizing the potentiality of light. These are rare and unfortunately the stage artist appreciative of the significations and emotional value of light and color is still rarer.

The equipment of the present stage consists of footlights, side-lights, border-lights, flood-lights, spot-lights, and much special apparatus. One of the severest criticisms of stage-lighting from an artistic point of view may be directed against the use of footlights for obtaining the dominant light. This is directed upward and the effect is an unnatural and even a grotesque modeling of the actors' features. The shadows produced are incongruous, for they are opposed to the other real and painted effects of light and shade. The only excuse for such lighting is that it is easily done and that proper lighting is difficult to obtain, owing to the fact that it involves a change in construction. By no means should the footlights be abandoned, for they would still be invaluable in obtaining diffused light even when the dominant light is directed from above the horizontal. In the present stage-lighting, in which the footlights generally predominate, the expressiveness of light is not satisfactory. Perhaps they are a necessary compromise, but inasmuch as their effect is unnatural they should not be accepted until it is thoroughly proved that ingenuity cannot eliminate the present defects.

The stage as a whole is a mobile picture in light, shade, and color with the addition of words and music. Excepting the latter, it is an expression of light worthy of the same care and consideration that the painting, which is also an expression of light, receives from the artist. The scenery and costumes should be considered in terms of the lighting effects because they are affected by changes in the color of the light. In fact, the author showed a number of years ago that by carefully relating the colors of the light with the colors used in painting the scenery, a complete change of scene can be obtained by merely changing the color of the light. Rather wonderful dissolving effects can be produced in this manner without shifting scenery. For example, a warm summer scene with trees in full foliage under a yellow light may be changed under a bluish light to a winter scene with ground covered with snow and trees barren of leaves. But before such accomplishments can be realized upon the stage, scientific knowledge must be available behind the scenes.

The art museum affords a multitude of opportunities for utilizing the expressiveness of light. This is more generally true of sculptured objects than of paintings because the latter may be treated as a whole. The artist almost invariably paints a picture by daylight and unless it is illuminated by daylight it is altered in appearance, that is, it becomes another picture. The great difference in the appearance of a painting under daylight and ordinary artificial light is quite startling, when demonstrated by means of apparatus in which the two effects may be rapidly alternated. Art museums are supposed to exhibit the works of artists and, therefore, no changes in these works should be tolerated if they can be avoided. The modern artificial-daylight lamps make it possible to illuminate galleries with light at night which approximates daylight. A further advantage of artificial light is that it may be easily controlled and a more satisfactory lighting may be obtained than with natural light. Considering the cost of daylight in museums and its disadvantages it appears possible that artificial daylight with its advantages may replace it eventually in the large galleries. If the works of artists are really prized for their appearance, the lighting of them is very important.

Sculpture is modeled by light and although it is impossible to ascertain the lighting under which the sculptor viewed his completed work with pride and satisfaction, it is possible to give the best consideration to its lighting in its final place of exhibition. The appearance of a sculpture depends upon the dominant direction of the light, the solid-angle subtended by the light-source (skylight, area of sky, etc.) and the amount of scattered light. The direction of dominant light determines the general direction of the shadows; the solid-angle of the light-source affects the character of the edges of the shadows; and the scattered light accounts for the brightness of the shadows. It should be obvious that variations of these factors affect the appearance or expression of three-dimensional objects. Therefore the position of a sculptured object with respect to the window or other skylight and the amount of light reflected from the surroundings are important. Visits to art museums with these factors in mind reveal a gross neglect in the lighting of objects of art which are supposed to appeal by virtue of their appearances, for they can arouse the emotions only through the doorway of vision.

A century ago mankind gave no thought to utilizing the expressive and impressive powers of light except in religious ceremonies. It was not practicable to utilize light from the feeble flames of those days in the elaborate manner necessary to draw upon these powers. Man was concerned with the more pressing needs. He wanted enough light to make the winter evenings endurable and the streets reasonably safe. The artists of those days saw the wonderful expressions of light exhibited by Nature, but they dared not dream of rivaling these with artificial light. To-day Nature surpasses man in the production of lighting effects only in magnitude. Man surpasses her artistically. In fact, the artist becomes a master only when he can improve upon her settings; when he is able by rare judgment in choosing and in eliminating and by skill and ingenuity to substitute a complete harmony for her incomplete and unsatisfactory reality. But everywhere Nature is the great teacher, for her world is full of an everchanging infinitude of expressions of light. Mankind needs only to study these with an attuned sensibility to be able eventually to play the music of light for those who are blessed with an esthetic sense.

XXIV

LIGHTING THE HOME

In the home artificial light exerts its influence upon every one. Without artificial lighting the family circle may not have become the important civilizing influence that it is to-day. Certainly civilized man now shudders at the thought of spending his evenings in the light of the fire upon the hearth or of a burning splinter.

The importance of artificial light is emphatically impressed upon the householder when he is forced temporarily to depend upon the primitive candle through the failure of the modern system of lighting. He flees from his home to that of his more fortunate neighbor, or he retires in his helplessness to awaken in the morning with a blessing for daylight. He cannot conceive of happiness and recreation in the homes of a century or two ago, when a few candles or an oil-lamp or two were the sole sources of light. But when the electric or gas service is again restored he relapses shortly into his former placid indifference toward the wonderfully efficient and adequate artificial light of the present age.

Until recently artificial light was costly and the householder in common with other users of light did not concern himself with the question of adequate and artistic lighting. His chief aim was to utilize as little as possible, for cost was always foremost in his mind. The development of the science of light-production has been so rapid during the past generation that adequate, efficient, and cheap artificial light finds mankind unconsciously viewing lighting with the same attitude as he displays toward his food and fuel bills. Another consequence of this rapid development is that mankind does not know how to extract the joy from modern artificial light. This is readily demonstrated by analyzing the lighting of middle-class homes.

The cost of light has been discussed in another chapter and it has been shown that it has decreased enormously in a century. It is now the most potential agency in the home when viewed from the standpoint of cost. The average householder pays less than twenty dollars per year for ever-ready light throughout his home. For about five cents per day the average family enjoys all the blessings of modern lighting, which is sufficient proof that cost is an insignificant item.

In order to simplify the discussion of lighting the home the terminology of electric-lighting will be used. The principles expounded apply as well to gas as to electricity, and owing to the ingenuity of the gas-lighting experts, the possibilities of gas-lighting are extensive despite its handicaps. There are some places in the home, such as the kitchen and basement, where lighting is purely utilitarian in the narrow sense, but in most of the rooms the esthetic or, more broadly, the psychological aspects of lighting should dominate. Pure utility is always a by-product of artistic lighting and furthermore, the lighting effects will be without glare when they satisfy all the demands of esthetics.

In dealing with lighting in the home the householder should concentrate his attention upon lighting effects. Unfortunately, he is not taught to do so, for everywhere he turns for help he finds the discussion directed toward fixtures and lamps instead of toward lighting effects. However, these are merely links in the chain from the meter to the eye. Lamps are of interest from the standpoint of quantity and quality of light, and fixtures are of importance chiefly as distributers of light. These details are merely means to an end and the end is the lighting effect. Of course, the fixtures are more important as objects than the wires because they are visible and should harmonize with the general decorative and architectural scheme.

The home is the theater of life full of various moods and occasions; hence the lighting of a home should be flexible. A degree of variety should be possible. Controls, wiring, outlets, and fixtures should conspire to provide this variety. At the present time the average householder does not give much attention to lighting until he purchases fixtures. It is probable that he thought of it when he laid out or approved the wiring, but usually he does not consider it seriously until he visits the fixture-dealer to purchase fixtures. And then unfortunately the fixture-dealer does not light his home; he does not sell the householder lighting-effects designed to meet the requirements of the particular home; he sells merely fixtures.

Unfortunately there are few fixtures available which have definite aims in lighting as demanded by the home. Of the great variety of fixtures available there are many artistic objects, but it is obvious that little attention is given to their design from the standpoint of lighting. That the fixture-dealer usually thinks of fixtures as objects and gives little or no thought to lighting effects is apparent from his conversation and from his display. He exhibits fixtures usually en masse and seldom attempts to illustrate the lighting effects produced in the room.

The foregoing criticisms are presented to emphasize the fact that throughout the field of lighting the great possibilities which have been opened by modern light-sources are not fully appreciated. The point at which to begin to design the lighting for a home is the wiring. Unfortunately this is too often done by a contractor who has given no special thought to the possibilities of lighting and to the requirements in wiring and switches necessary in order to realize them. At this point the householder should attempt to form an opinion as to the relative values. Is artificial lighting important enough to warrant an expenditure of two per cent. of the total investment in the home and its furnishings? The answer will depend upon the extent to which artificial light is appreciated. It appears that four or five per cent. is not too much if it is admitted that the artificial lighting system ranks next to the heating plant in importance and that these two are the most important features of an interior of a residence. A switch or a baseboard outlet costs an insignificant sum but either may pay for itself many times in the course of a few years through its utility or convenience.

It appears best to take up this subject room by room because the requirements vary considerably, but in order to be specific in the discussions, a middle-class home will be chosen. The more important rooms will be treated first and various simple details will be touched upon because, after all, the proper lighting of a home is realized by attention to small details.

The living-room is the scene of many functions. It serves at times for the quiet gathering of the family, each member devoted to reading. At another time it may contain a happy company engaged at cards or in conversation. The lighting requirements vary from a spot or two of light to a flood of light. Excepting in the small living-rooms there does not appear to be a single good reason for a ceiling fixture. It is nearly always in the field of vision when occupants are engaged in conversation, and for reading purposes the portable lamp of satisfactory design has no rival. Wall brackets cannot supply general lighting without being too bright for comfort. If they are heavily shaded they may still emit plenty of light upward, but the adjacent spots on the walls or ceiling will generally be too bright. Wall brackets may be beautiful ornaments and decorative spots of light and have a right to exist as such, but they cannot be safely depended upon for adequate general lighting on those occasions which demand such lighting.

As a general principle, it is well to visualize the furniture in the room when looking at the architect's drawings and it is advantageous even to cut out pieces of paper representing the furniture in scale. By placing these on the drawings the furnished room is readily visualized and the locations of baseboard outlets become evident. It appears that the best method of lighting a living-room is by means of decorative portable lamps. Such lamps are really lighting-furniture, for they aid in decorating and in furnishing the room at all times. A number of these lamps in the living-room insures great flexibility in the lighting, and the light may be kept localized if desired so that the room is restful. A room whose ceiling and walls are brilliantly illuminated is not so comfortable for long periods as one in which these areas are dimly lighted. Furthermore, the latter is more conducive to reading and to other efforts at concentration. The furniture may be readily shifted as desired and the portable lamps may be rearranged.

Such lighting serves all the purposes of the living-room excepting those requiring a flood of light, but it is easy to conceal elaborate lighting mechanisms underneath the shades of portable lamps. Several types of portable lamps are available which supply an indirect component as well as direct light. The former illuminates the ceiling with a flood of light without any discomforting glare. Such a lighting-unit is one of the most satisfactory for the home, for two distinct effects and a combination of these introduce a desirable element of variety into the lighting. Not less than four and preferably six baseboard outlets should be provided in a living-room of moderate size. One outlet on the mantel is also to be desired for connecting decorative candlesticks, and brackets above the fireplace are of ornamental value. Although the absence of ceiling fixtures improves the appearance of the room, wiring may be provided for ceiling outlets in new houses as a matter of insurance against the possible needs of the future. When ceiling fixtures are not used, switches may be provided for the mantel brackets or certain baseboard outlets in order that light may be had upon entering the room.

The merits of a portable lamp may be ascertained before purchasing by actual demonstration. Some of them are not satisfactory for reading-lamps, owing to the shape of the shade or to the position of the lamps. The utility of a table lamp may be determined by placing it upon a table and noting the spread of light while seated in a chair beside it. A floor lamp may also be tested very easily. A miniature floor lamp about four feet in height with an appropriate shade provides an excellent lamp for reading purposes because it may be placed by the side of a chair or moved about independent of other furniture. A tall floor lamp often serves for lighting the piano, but small piano lamps may be found which are decorative as well as serviceable in illuminating the music without glare.

The dining-room presents an entirely different problem for the setting is very definite. The dining-table is the most important area in the room and it should be the most brilliantly illuminated area in the room. A demonstration of this point is thoroughly convincing. The decorator who designs wall brackets for the dining-room is interested in beautiful objects of art and not in a proper lighting effect. The fixture-dealer, having fixtures to sell and not recognizing that he could fill a crying need as a lighting specialist, is as likely to sell a semi-indirect or an indirect lighting fixture as he is to provide a properly balanced lighting effect with the table brightly illuminated. The indirect and semi-indirect units illuminate the ceiling predominantly with the result that this bright area distracts attention from the table. A brightly illuminated table holds the attention of the diners. Light attracts and a semi-darkness over the remainder of the room crowds in with a result that is far more satisfactory than that of a dining-room flooded with light.

The old-fashioned dome which hung over the dining-table has served well, for it illuminated the table and left the remainder of the room dimly lighted. But its wide aperture made it necessary to suspend it rather low in order that the lamps within should not be visible. It is an obtrusive fixture and despite its excellent lighting effect, it went out of style. But satisfactory lighting principles never become antiquated, and as taste in fixtures changes the principles may be retained in new fixtures. Modern domes are available which are excellent for the dining-room if the lamps are well concealed. The so-called showers are satisfactory if the shades are dense and of such shape as to conceal the lamps from the eyes. Various modifications readily suggest themselves to the alert fixture-designer. Even the housewife can do much with silk shades when the principle of lighting the dining-table is understood. The so-called candelabra have been sold extensively for dining-rooms and they are fairly satisfactory if equipped with shades which reflect much of the light downward. Semi-indirect and indirect fixtures have many applications in lighting, but they do not provide the proper effect for a dining-room.

It is easy to make a special fixture which will send a component of light downward to the table and will permit a small amount of diffused light to the ceiling and walls. If a daylight lamp is used for the direct component, the table will appear very beautiful. Under this light the linen and china are white, flowers and decorations on the china appear in their full colors, the silver is attractive, and the various color-harmonies such as butter, paprika, and baked potato are enticing. This is an excellent place for a daylight lamp if diffused light illuminating the remainder of the room and the faces of the diners is of a warm tone obtained by warm yellow lamps or by filtering these components of the light through orange shades. The ceiling fixture should be provided with two circuits and switches. In some cases it is easy to provide a dangling plug for connecting such electric equipment as a toaster, percolator, or candlesticks. Two candlesticks are effective on the buffet, but usually the smallest normal-voltage lamps available give too much light. Miniature lamps may be used with a small transformer, or two regular lamps may be connected in series. At least two baseboard outlets are convenient.

The foregoing deals with the more or less essential lighting of a dining-room, but there are various practicable additional lighting effects which add much charm to certain occasions. Colored light of low intensity obtained from a cove or from "flower-boxes" fastened upon the wall is very pleasing. If a cove is provided around the room, two circuits containing orange and blue lamps respectively will supply two colors widely differing in effect. By mixing the two a beautiful rose tint may be obtained. This equipment has been installed with much satisfaction. A simpler method of obtaining a similar effect is to use imitation flower-boxes plugged into wall outlets. Artificial foliage adds to the charm of these boxes. The colored light is merely to add another effect on special occasions and its intensity should never be high. In the dining-room such unusual effects are not out of place and they need not be garish.

The sun-room partakes of the characteristics of the living-room to some extent, but, it being smaller, a semi-indirect fixture may be satisfactory for general illumination. However, a portable lamp which supplies an indirect component of light besides the direct light serves admirably for reading as well as for flooding the room with light when necessary. Two or three baseboard outlets are desirable for attaching decorative or even purely utilitarian lamps. The sun-room is an excellent place for utilizing "flower-box" fixtures decorated with artificial foliage. In fact, a central fixture may assume the appearance of a "hanging basket" of foliage. The library and den offer no problems differing from those already discussed in the living-room. A careful consideration of the disposition of the furniture will reveal the best positions for the outlets. In a small library wall brackets may serve as decorative spots of light and if the shades are pendent they may serve as lamps for reading purposes. In both these rooms an excellent reading-lamp is desired, but it may be decorative as well. Wall outlets may be desired for decorative portable lamps upon the bookcases.

The sleeping-room, which commonly is also a dressing-room, often exhibits the errors of a lack of foresight in lighting. In most rooms of this character there is one best arrangement of furniture and if this is determined it is easy to ascertain where the windows and outlets should be located. The windows may usually be arranged for twin beds as well as for a single one with obvious advantages of flexibility in arrangement. With the position of the bureau determined it is easy to locate outlets for two wall brackets, one on each side, about sixty-six inches above the floor and about five feet apart. When the brackets are equipped with dense upright shades, the figure before the mirror is well illuminated without glare and sufficient light reaches the ceiling to illuminate the whole room.

A baseboard outlet should be available for small portable lamps which may be used upon the bureau or for electric heating devices. The same is true for the dressing-table; indeed, two small decorative lamps on the table serve better than high wall brackets owing to the fact that the user is seated. A baseboard outlet near the head of the bed or between the beds is convenient for a reading-lamp and for other purposes. An outlet in the center of the ceiling controlled by a convenient switch may be installed on building, as insurance against future needs or desires. But a single lighting-unit in the center of the ceiling does not serve adequately the needs at the bureau and dressing-table. In fact, two wall brackets properly located with respect to the bureau afford a lighting much superior for all purposes in the bedroom to that produced by a ceiling fixture.

In the bath-room the principal problem is to illuminate the person, especially the face, before the mirror. Many mistakes are made at this point, despite the simplicity of the solution. In order to see the image of an object in a mirror, the object must be illuminated. It is best to do this in a straightforward manner by means of a small lighting-unit on each side of the mirror at a height of five feet. Both sides of the face will be well illuminated and the light-sources are low enough to eliminate objectionable shadows. The units may be merely pull-chain sockets containing frosted or opal lamps. A center bracket or a single unit suspended from the ceiling is not as satisfactory as the two brackets. These afford enough light for the entire bath-room. A baseboard or wall outlet is convenient for connecting a heater, curling-iron, and other electrically heated devices.

The sewing-room, which in the middle-class home is usually a small room, is sometimes used as a bedroom. A ceiling fixture will supply adequate general lighting, but a baseboard outlet should be available for a short floor lamp or a table lamp for sewing purposes. An intense local light is necessary for this occupation, which severely taxes the eyes. A so-called daylight lamp serves very well in this case.

Towering shafts of light defy the darkness and thousands of lighted windows symbolize man's successful struggle against nature]

In the kitchen the wall brackets are easily located after the positions of the range, work-table, sink, etc., are determined. A bracket for each is advisable unless they are so located that one will serve two purposes. It is customary to have a combination fixture for gas and electricity. This is often suspended from the center of the ceiling, but inasmuch as the gas-light cannot be close to the ceiling, the fixture extends so far downward as to become a nuisance. Furthermore, a light-source hung low from the center of the ceiling is in such a position that the worker in the kitchen usually works in his shadow. If a ceiling outlet is used it should be an electrical socket at the ceiling. The combination fixture is best placed on the wall as a bracket. The so-called daylight lamps are valuable in the kitchen.

In the basement a generous supply of ceiling outlets adds much to the satisfaction of a basement. One in each locker, one before the furnace, and a large daylight lamp above but to one side of the laundry trays are worth many times their cost. Furthermore, a wall socket for the electric iron and washing-machine is a convenience very much appreciated.

In the stairways convenient three-way switches for each of the ceiling fixtures represents the best practice. A baseboard outlet in the upper hall affords a connection for a decorative lamp and pays for itself many times as a place to attach the vacuum-cleaner from which all the rooms on that floor may be served. In vestibules and on porches ceiling fixtures controlled by means of convenient switches are satisfactory. The entrance hall may be made to express hospitality by means of lighting which should be adequate and artistic.

An adequate supply of outlets and wall switches is not costly and they pay generous dividends. With a scanty supply of these, the possibilities of lighting are very much curtailed. There is nothing intricate about locating switches and outlets, so the householder may do this himself, or he may view critically the plans as submitted. The chief difficulties are to throw aside his indifference and to readjust his ideas and values. It may be confidently stated that the possibilities of lighting far outrank most of the features which contribute to the cost of a house and of its furnishings.

After considering the requirements and decorative schemes of the various rooms the householder should be competent to judge the appropriateness of the lighting effects obtained from fixtures which the dealer displays, but he should insist upon a demonstration. If the dealer is not equipped with a room for this purpose, he should be asked to demonstrate in the rooms to be lighted. If the fixture-dealer does not realize that he should be selling lighting effects, the householder should make him understand that lighting effects are of primary importance and the fixtures themselves are of secondary interest in most cases. The unused outlets that have been installed for possible future needs may be sealed in plastering if the positions are marked so that they may be found when desired.

An advantage of portable lamps is that they may be taken away on moving. In fact, when lighting is eventually considered a powerful decorative medium, as it should be, it is probable that fixtures will be personal property attached to ceiling, wall, and floor outlets by means of plugs.

A variety of incandescent lamps are available. For the home, opal, frosted, or bowl-frosted lamps are usually more satisfactory than clear lamps. Bare filaments should not be visible, for they not only cause discomfort and eye-strain but they spoil what might otherwise be an artistic effect. Lamps with diffusing bulbs do much toward eliminating harsh shadows cast by the edges of the shades, by the chains of the fixtures, etc. These lamps are available in many shapes and sizes and the householder should make a record of voltage, wattage, and shape of the lamps which he finds satisfactory in the various fixtures. The Mazda daylight lamp has several places in the home and the Mazda white-glass and other high-efficiency lamps supply many needs better than the vacuum lamps. In brackets and other purely decorative lighting-units small frosted lamps are usually the most satisfactory. There is a general desire for the warm yellowish light of the candle-flame, and this may be obtained by a tinted shade but usually more satisfactorily by means of a tinted lamp.

The householder will find it interesting to become intimate with lighting, for it can serve him well. The housewife will often find much interest in making shades of textiles and of parchment. Charming glassware in appropriate tints and painted designs is available for all rooms. In the bedchamber and the nursery some of these painted designs are exceedingly effective. Fixtures should shield the lamps from the eyes, and the diffusing media whether glass or textile should be dense enough to prevent glare. No fixture can be beautiful and no lighting effect can be artistic if glare is present. If the architect and the householder will realize that light is a medium comparable with the decorator's media, better lighting will result. Light has the great advantage of being mobile and with adequate outlets and controls supplemented by fixtures from which different effects are available, the householder will find in lighting one of the most fruitful sources of interest and pleasure. It can do much toward expressing the taste of the householder or if neglected it can undo much of the effect of excellent decoration and furnishing. Artificial lighting, softly diffused and properly localized, is one of the most important factors in making a house a home.

XXV

LIGHTING--A FINE ART?

In the preceding chapters the progress of light has been sketched from its obscure infancy to its vigorous youth of the present time. It has been seen that progress was slow until the beginning of the nineteenth century, after which it began to gain momentum until the present century has witnessed tremendous advances. Until the latter part of the nineteenth century artificial light was considered an expensive utility, but as modern lamps appeared which supplied adequate light at reasonable cost attention began to be centered upon utilization, and the lighting engineer was born. Gradually it is being realized that artificial light is no longer a luxury, that it may be used in great quantity, and that it may be directed, diffused, and altered in color as desired. Although the potentiality of light has been barely drawn upon, the present usages surpass the most extravagant dreams of civilized beings a half-century ago. Mere light of that time was changed into more light as gas-lighting developed, and more light has increased to adequate light of the present time through the work of scientists.

It is apparent that a sudden enforced reversion to the primitive flames of fifty years ago would paralyze many activities. Much of interest and beauty would be blotted out of this brilliant, pulsating, productive age. It is startling to note that almost the entire progress in artificial lighting has taken place during the past hundred years and that most of it has been crowded into the latter part of this period. In fact, its development since it began in earnest has gone forward with ever-increasing momentum. On viewing the wonders of modern artificial lighting on every hand it is not difficult to muster the courage necessary to venture into its future.

The lighting engineer has been a natural evolution of the present age, for the economic aspects of lighting have demanded attention. He is increasing the safety, efficiency, and happiness of mankind and civilization is beginning to feel his influence economically. However, with the advent of adequate, efficient, and controllable light, the potentiality of light as an artistic medium may be drawn upon and the lighting artist with a deep insight into the possibilities of artificial light now has his opportunity. But the artist who believes that a new art may be evolved to perfection in a few years is doomed to disappointment, for it is necessary only to view retrospectively such arts as painting and music to be convinced that understanding and appreciation develop slowly through centuries of experiment and contact.

Will lighting ever become a fine art? Will it ever be able alone to arouse emotional man as do the fine arts? Are the powers of light sufficiently great to enthrall mankind without the aid of form, music, action, or spoken words? It is safer to answer "yes" than "no" to these questions. Painting has reached a high place as an art and this art is the expressiveness of secondary or reflected light reinforced by imitation forms, which by a combination of light and drawing comprise the "subjects." A painting is a momentary expression of light, a cross-section of something mobile, such as nature, thought, or action. Light has the essential qualifications of painting with the advantages of a greater range of brightness, of greater purity of colors, and the great potentiality of mobility. If lighting becomes a fine art it will doubtless be related to painting somewhat in the same manner that architecture is akin to sculpture. With the introduction of mobility it will borrow something from the arts of succession and especially from music.

The art of lighting in its present infancy is leaning upon established arts, just as the infant learns to walk alone by first depending upon support. The use of color in painting developed slowly, being supported for centuries by the strength of drawing or subject. The landscapes of a century ago were dull, for color was employed hesitatingly and sparingly. The colors in the portraits of the past merely represented the gorgeous dress of bygone days. But the painter of the present shows that color is beginning to be used for itself and that the painter is no longer hesitant concerning its power to go hand in hand with drawing. Drafting and coloring are now in partnership, the former having given up guardianship when the latter reached maturity.

Lighting is now an accompaniment of the drama, of the dance, of architecture, of decoration, and of music. It has been a background or a part of the "atmosphere" excepting occasionally when some one with imagination and daring has given it the leading role. Even in its infancy it has on occasions performed admirably almost without any aid. The bursting rocket, the marvelous effects at the Panama-Pacific Exposition, and some of the exhibitions on the theatrical stage are glimpses of the potentiality of light. To fall back upon the terminology of music, these may be glimmerings of light-symphonies.

Harmony is simultaneity and a painting in this respect is a chord--a momentary expression fixed in material media. A melody of light requires succession just as the melody in music. The restless colors of the opal comprise a light melody like the songs of birds. The gorgeous splendor of the sunset compares in magnitude and in its various moods with the symphony orchestra and its powers. Throughout nature are to be found gentle chords, beautiful melodies and powerful symphonies of light and this music of light exhibits the complexity and structure analogous to music. There is no physical relation between music, poetry, and light, but it is easy to lean upon the established terminology for purposes of discussion. Those who would build color-music identical to sound music are making the mistake of starting with a physical foundation instead of basing the art of light-expression upon psychological effects of light. In other words, a relation between light and music can exist only in the psychological realm.

These melodies and symphonies of light in nature are admittedly pleasing or impressive as the case may be, but are they as appealing as music, poetry, painting, or sculpture? The consensus of opinion of a large group of average persons might indicate a negative reply, but the combined opinion of this group is not so valuable as the opinion of a colorist or of an artist who has sensed the wonders of light. The unprejudiced opinion of artists is that light is a powerfully expressive and impressive medium. The psychologist will likely state that the emotive value of light or color is not comparable to the appeal of an excellent dinner or of many other commonplace things. But he has experimented only with single colors or with simple patterns and his subjects are selected more or less at random from the multitude. What would be his conclusion if he examined painters and others who have developed their sensibilities to a deep appreciation of light and color? It is certain that the painter who picks up a purple petal fallen from a rose and places it upon a green leaf is as thrilled by the powerful vibrant color-chord as the musician who hears an exquisite harmony of sounds.

Music has been presented to civilized mankind in an organized manner for ages and the fundamental physical basis of modern music is a thousand years old. Would the primitive savage appreciate the modern symphony orchestra? Even the majority of civilized beings prefer the modern ragtime or jazz to the exquisite art of the symphony. An appreciation of the opera and the symphony is reached by educational methods extending over long periods. An appreciation of the expressiveness of light cannot be expected to be realized by any short-cut. Most persons to-day enjoy the melodramatic "movie" more than the drama and relatively few experience the deep appeal of the fine arts. Surely the symphony of light cannot be justly condemned because of a lack of appreciation and understanding of it, for it has not been introduced to the public. Furthermore, the expressiveness of music is still indefinite at best despite the many centuries of experimenting on the part of musicians.

If poetry is to be believed, the symphonies of light as rendered by nature in the sunsets, in the aurora borealis, and in other sky-effects of great magnitude have deeply impressed the poet. If his descriptions are to be accepted at their face-value, the melodies of light rendered in the precious stone, in the ice-crystal, and in the iridescence of bird-plumage please his finer sensibilities. If he is sincere, mobile light is a seductive agency.

The painter has contributed little of direct value in developing the music of light. He is concerned with an instantaneous expression. He waits for it patiently and, while waiting, learns to appreciate the fickleness of mood in nature, but when he fixes one of these moods he has contributed very little to the art of mobile light. Unfortunately the art schools teach the student little or nothing pertaining to color for color's sake. When the student is capable of drawing fairly well and is acquainted with a few stereotyped principles of color-harmony he is sent forth to follow in the footsteps of past masters. He may be seen at the art museum faithfully copying a famous painting or out in the fields stalking a tree with the hopes of an embryo Corot. The world moves and has only a position in the rank and file for imitators. Occasionally an artist goes to work with a vim and indulges in research, thereby demonstrating originality in two respects. Painting is just as much a field for research as light-production.

Recently experiments are being made in the production of color-harmonies devoid of form. Surely there is a field for pure color-composition and this the field of the painter which leads toward the art of mobile light. Many of the formless paintings of the present day which pass under the banner of this ism or that are merely experiments in the expressiveness of light. Being formless, they are devoid of subject in the ordinary sense and cannot be more or less than a fixed expression of light. Naturally they have received much criticism and have been ridiculed, but they can expect nothing else until they are understood. They cannot be understood until mankind learns their language and then they must be understandable. In other words, there are impostors gathered around the sincere research-artist because the former have neither the ability to paint for a living nor the inclination to forsake the comparative safety of the mystery of art for the practical world where their measure would be quickly taken. This army of camp-followers will not advance the art of mobile light, but the sincere seekers after the principles of light-expression who form the foundation of the various isms may contribute much.

The painter will always be available with his finer sensibility to appreciate and to aid in developing the art of mobile light, but his direct contribution appears most likely to come from the present chaos of experiments in pure color-composition, in the psychology of light, or, more broadly, in the expressiveness of light. The decorator and the designer of gowns and costumes do not arrogate to themselves the name "artist," but they are daily creating something which is leading toward a fuller appreciation of the expressiveness of light. If they do not contribute directly to the development of the art of mobile light, they are at least aiding in developing what may eventually be an appreciative public.

The artist paints a "still-life," the decorator creates a color-harmony of abstract or conventional forms, and the costumer produces a color-composition in textiles. The decorator and costumer approach closer to pure color-composition than the artist in his still-life. The latter is a grouping of objects primarily for their color-notes. Why bother with a banana when a yellow-note is desired? Why utilize the abstract or conventional forms of the decorator? Why not follow this lead further to the less definite forms employed by the costumer? Why not eliminate form even more completely? This is an important point and an interesting lead, for to become rid of form has been one of the perplexing problems encountered by those who have dreamed of an art of mobile light.

The painter who uses line and color imitatively has perhaps acquired skill in depicting objects and more or less appreciation of the beautiful. But if he is to be creative and to produce a higher art he must be able to use line and color without reference to objects. He thus may aid in the development of an abstract art which is the higher art and at the same time aid in educating the public to appreciate pure color-harmonies. From these momentary expressions of light and from the experience gained, the mobile colorist would receive material aid and his productions would be viewed by a more receptive audience or rather "optience" as it may be called. The development of taste for abstract art is needed in order that the art of mobile light may develop and, incidentally, an appreciation of the abstract in art is needed in all arts.

Science has contributed much by way of clearing the decks. It has produced the light-sources and the apparatus for controlling light. It has analyzed the physical aspects of color-mixture and has accumulated extensive data pertaining to color-vision. It has pointed out pitfalls and during recent years has been delving further by investigating the psychology of light and color. The latter field is looked to for valuable information, but, after all, there is one way of making progress in the absence of data and that is to make attempts at the production of impressive effects of mobile light. Some of these have been made, but unfortunately they have been heralded as finished products.

Perhaps the most general mistake made is in relating sounds and colors by stressing a mere analogy too far. Notwithstanding the vibratory nature of the propagation of sound and light, this is no reason for stressing a helpful analogy. After all it is the psychological effect that is of importance and it is absurd to attribute any connection between light-waves and sound-waves based upon a relation of physical quantities. No space will be given to such a relation because it is so absurdly superficial; however, the language of music will be borrowed with the understanding that no relation is assumed.

A few facts pertaining to vision will indicate the trend of developments necessary in the presentation of mobile light. The visual process synthesizes colors and at this point departs widely from the auditory process. The sensation of white may be due to the synthesis of all the spectral colors in the proportions in which they exist in noon sunlight or it may be due to the synthesis of proper proportions of yellow and blue, of red, green, and blue, of purple and green, and a vast array of other combinations. A mixture of red and green lights may produce an exact match for a pure yellow. Thus it is seen that the mixture of lights will cause some difficulty. For example, the components of a musical chord may be picked out one by one by the trained ear, but if two or more colored lights are mixed they are merged completely and the resultant color is generally quite different from any of the components. In music of light, the components of color-chords must be kept separated, for if they are intermingled like those of musical chords they are indistinguishable. Therefore, the elements of harmony in mobile light must be introduced by giving the components different spatial positions.

The visual process is more sluggish than the auditory process; that is, lights must succeed each other less rapidly than musical notes if they are to be distinguished separately. The ear can follow the most rapid execution of musical passages, but there is a tendency for colors to blend if they follow one another rapidly. This critical frequency or rate at which successive colors blend decreases with the brightness of the components. If red and green are alternated at a rate exceeding the critical frequency, a sensation of yellow will result; that is, neither component will be distinguishable and a steady yellow or a yellow of flickering brightness will be seen. The hues blend at a lower frequency than the brightness components of colors; hence there may be a blend of color which still flickers in brightness. Many weird results may be obtained by varying the rate of succession of colors. If this rate is so low that the colors do not tend to merge, they are much enriched by successive contrast. It is known that juxtaposed colors generally enrich one another and this phenomenon is known as simultaneous contrast. Successive contrast causes a similar effect of heightened color.

An effect analogous to dynamic contrast in music may be obtained with mobile light by varying the intensity of the light or possibly the area. Melody may be simply obtained by mere succession of lights. Tone-quality has an analogy in the variation of the purity of color. For example, a given spectral hue may be converted into a large family of tints by the addition of various amounts of white light. Rhythm is as easily applied to light as to music, to poetry, to pattern, or to the dance, but in mobile lights its limitations already have been suggested. However, it is bound to play an important part in the art of mobile light because rhythmic experiences are much more agreeable than those which are non-rhythmic. Rhythm abounds everywhere and nothing so stirs mankind from the lowliest savage to the highly cultivated being as rhythmic sequences.

Many psychological effects of light have been recorded from experiment and observation and affective values of light have been established in various other byways. It is possible that the degree of pleasure experienced by most persons on viewing a color-harmony or the delightful color-melody of a sunlit opal may be less than that experienced on listening to the rendition of music. However, if this were true it would offer no discouragement, because absolute values play a small part in life. Two events when directly compared apparently may differ enormously in their ability to arouse emotions, but the human organism is so adaptive that each in its proper environment may powerfully affect the emotions. For example, those who have sported in antics in the heights of cloudland or have stormed the enemy's trench are still capable of enjoying a sunset or the call of a bird to its mate at dusk. The wonderful adaptability of the inner being is the salvation of art as well as of life.

In the rendition of mobile light it is fair to give the medium every advantage. Sometimes this means to eliminate competitors and sometimes it means to remove handicaps. On the stage light has had competitors which are better understood. For example, in the drama words and action are easily understood, and regardless of the effectiveness of light it would not receive much credit for the emotive value of the production. In the wonderful harmony of music, dance, and light in certain recent exhibitions, the dance and music overpowered the effects of lights because they speak familiar languages.

A community song-festival

ARTIFICIAL LIGHT IN COMMUNITY AFFAIRS]

Artificial light not only reveals the beauty of decoration and architecture but enthralls mankind with its own unlimited powers]

A number of attempts have been made to utilize light as an accompaniment of music and some of them on a small scale have been sincere and creditable, but a much-heralded exhibition on a large scale a few years ago was not the product of deep thought and sincere effort. For example, colored lights thrown upon a screen having an area of perhaps twenty square feet were expected to compete with a symphony orchestra in Carnegie Hall. The music reached the most distant auditor in sufficient volume, but the lighting effect dwindled to insignificance. Without entering into certain details which condemned the exhibition in advance, the method of rendition of the light-accompaniment revealed a lack of appreciation of the problems involved on the part of those responsible.

Incidentally, it has been shown that the composer of this particular musical selection with its light accompaniment was psychologically abnormal; that is, he was affected with colored audition. It is not yet established to what extent normal persons are similarly affected by light and color. Certainly there is no similarity among the abnormal and none between the abnormal and normal.

If light is to be used as an accompaniment to music, it must be given an opportunity to supply "atmosphere." This it cannot do if confined to an insignificant spot; it must be given extensity. Furthermore, by the use of diaphanous hangings, form will be minimized and the evanescent effects surely can be charming. But finally the lighting effects must fill the field of vision just as the music "fills the field of audition" in order to be effective. There are fundamental objections to the use of mobile light as an accompaniment to music and therefore the future of the art of mobile light must not be allowed to rest upon its success with music. If it progresses through its relation with music, so much is gained; if not, the relation may be broken for music is quite capable of standing alone.

There is a tendency on the part of some revolutionary stage artists to give to lighting an emotional part in the play, or, in other words, to utilize lighting in obtaining the proper mood for the action of the play. Color and purely pictorial effect are the dominant notes of some of them. All of these modern stage-artists are abandoning the intricately realistic setting, and, as a consequence, light is enjoying a greater opportunity. In the more common and shallow theatrical production, lighting and color effects have many times saved the day, and, although these effects are not of the deeper emotional type, they may add a spectacular beauty which brings applause where the singing is mediocre and the comedy isn't comedy. The potentiality of lighting effects for the stage has been barely drawn upon, but as the expressiveness of light is more and more utilized on the stage, the art of mobile light will be advanced just so much more. Light, color, and darkness have many emotional suggestions which are easily understood and utilized, but the blending of mobile light with the action is difficult because its language is only faintly understood.

It is futile to attempt to describe a future composition of mobile light. Certainly there is an extensive variety of possibilities. A sunset may be compressed into minutes or an opalescent sky may be a motif. Varying intensities of a single hue or of allied hues may serve as a gentle melody. Realistic effects may be introduced. The expressiveness of individual colors may be taken as a basis for constructing the various motifs. These may be woven into melody in which rhythm both in time and in intensity may be introduced. Action may be easily suggested and the number of different colors, in a broad sense, which are visible is comparable to the audible tones. Shading is as easily accomplished as in music and the development of this art need not be inhibited by a lack of mechanical devices and light-sources. The tools will be forthcoming if the conscientious artist requests them.

Whatever the future of the art of mobile light may be, it is certain that the utilization of the expressiveness of light has barely begun. It may be that light-music must pass through the "ragtime" stage of fireworks and musical-revue color-effects. If so, it is gratifying to know that it is on its way. Certainly it has already served on a higher level in some of the artistic lighting effects in which mobility has featured to some extent.

If the art does not develop rapidly it will be merely following the course of other arts. A vast amount of experimenting will be necessary and artists and public alike must learn. But if it ever does develop to the level of a fine art its only rival will be music, because the latter is the only other abstract art. Material civilization has progressed far and artificial light has been a powerful influence. May it not be true that artificial light will be responsible for the development of spiritual civilization to its highest level? If mobile light becomes a fine art, it will be man's most abstract achievement in art and it may be incomparably finer and more ethereal than music. If this is realized, artificial light in every sense may well deserve to be known as the torch of civilization.

READING REFERENCES

No attempt will be made to give a pretentious bibliography of the literature pertaining to the various aspects of artificial lighting, for there are many articles widely scattered through many journals. The Transactions of the Illuminating Engineering Society afford the most fruitful source of further information; the Illuminating Engineer (London), contains much of interest; and Zeitschrift Beleuchtungswesen deals with lighting in Germany. H. R. D'Allemagne has compiled an elaborate "Historie du Luminaire" which is profusely illustrated, and L. von Benesch in his "Beleuchtungswesen" has presented many elaborate charts. In both these volumes lighting devices and fixtures from the early primitive ones to those of the nineteenth century are illustrated. A few of the latest books on lighting, in the English language, are "The Art of Illumination," by Bell; "Modern Illuminants and Illuminating Engineering," by Gaster and Dow; "Radiation, Light and Illumination," by Steinmetz; "The Lighting Art," by Luckiesh; "Illuminating Engineering Practice," consisting of a course of lectures presented by various experts under the joint auspices of the University of Pennsylvania and the Illuminating Engineering Society; "Lectures on Illuminating Engineering," comprising a series of lectures presented under the joint auspices of Johns Hopkins University and the Illuminating Engineering Society; and "The Range of Electric Searchlight Projectors," by Rey; "The Electric Arc," by Mrs. Ayrton; "Electric Arc Lamps," by Zeidler and Lustgarten, and "The Electric Arc," by Child treat the scientific and technical aspects of the arc. G. B. Barham has furnished a book on "The Development of the Incandescent Electric Lamp." "Color and Its Applications," and "Light and Shade and Their Applications," are two books by Luckiesh which deal with lighting from unique points of view. "The Language of Color," by Luckiesh, aims to present what is definitely known regarding the expressiveness and impressiveness of color. W. P. Gerhard has supplied a volume on "The American Practice of Gaspiping and Gas Lighting in Buildings," and Leeds and Butterfield one on "Acetylene." A recent book in French by V. Trudelle treats "Lumiere Electrique et ses differentes Applications au Theatre." Many books treat of photometry, power-plants, etc., but these are omitted because they deal with phases of light which have not been discussed in the present volume. "Light Energy," by Cleaves, is a large volume devoted to light-therapy, germicidal action of radiant energy, etc. References to individual articles will often be found in the various indexes of publications.

THE END